I0045031

Albrecht von Graefe, Hasket Derby

Clinical Lectures on Amblyopia and Amaurosis

Albrecht von Graefe, Hasket Derby

Clinical Lectures on Amblyopia and Amaurosis

ISBN/EAN: 9783337840327

Printed in Europe, USA, Canada, Australia, Japan

Cover: Foto ©berggeist007 / pixelio.de

More available books at **www.hansebooks.com**

CLINICAL LECTURES

BY

PROFESSOR A. VON GRAEFE,

ON

AMBLYOPIA AND AMAUROSIS,

AND THE

EXTRACTION OF CATARACT.

Translated from the German

By HASKET DERBY, M.D.

Surgeon to the Massachusetts Charitable Eye and Ear Infirmary ; Member of the Heidelberg
Ophthalmologische Gesellschaft ; Member of the American Ophthalmological
Society, &c. &c.

BOSTON:

DAVID CLAPP & SON, PRINTERS.....334 WASHINGTON ST.

1866.

PREFACE.

THIS translation was made during the past winter for the Boston Medical and Surgical Journal, and is now collected from its pages. Its publication is undertaken for the two-fold object of introducing Albrecht von Graefe to the American medical public as a clinical teacher, and of exhibiting the progress which has been made in the exploration of one of the most obscure departments of ophthalmic science, and for which we are indebted to his genius and industry.

In his first work on Astigmatism, published in 1862, Donders relates a case of Myopia, complicated seemingly with an amount of Amblyopia too great to be accounted for by the moderate near-sightedness. Local depletion and other treatment worked no change. Three years later, thanks to the progress of science in the interval, the difficulty was found to be dependent on Astigmatism, and was relieved by appropriate glasses. The illustrious professor thus comments on the case. "This instance is one of thousands where Astigmatism has been confounded with and treated as Amblyopia. The tortures the patient suffered from the severe and futile treatment were compensated for by the indescribable joy he experienced in finding his vision improved for every distance by appropriate glasses;" and in another place he says, "this discovery goes another step to contract the domain of as yet unexplained Amblyopia." Efforts for its relief were however not to stop here. The work has been pushed on with equal ardor by von Graefe, and those who have the patience to master his close and minute reasoning will find many difficulties that had previously beset their path in the treatment of these affections, explained and cleared away.

With reference to the series of clinical remarks on the extraction of cataract which form the conclusion of this pamphlet, and which, as a specimen of clear, precise and comprehensive clinical instruction, the translator believes to be without a parallel in ophthalmic literature, it may be well to remark that their value is in no wise diminished by the late researches of von Graefe on modified linear extraction and his present frequent adoption of the method. Till vastly more extended statistics have proved the superior safety of another procedure, flap extraction must remain the rule in a large proportion of cases and for the majority of practitioners.

The lectures on Amblyopia and Amaurosis were compiled and reported by Dr. Engelhardt, and were translated from the Klinische Monatsblätter für Augenheilkunde for 1865. Those on the extraction of cataract were taken from the volume of the same magazine for 1863.

6 BEACON STREET, BOSTON,
  May 25, 1866.

AMBLYOPIA AND AMAUROSIS.

In cases of Amblyopia,* three things aid us in general in arriving at our conclusions. First, the *functional state* of the eye, carefully considered; second, the *appearance of the papilla*; third, the *manner in which the affection has become developed*.

As regards the functional derangements, it is not to be denied that the amount of impairment of central vision is of great importance; although the prognosis as regards possible blindness would depend yet more on an accurate investigation of the limits of the field of vision and of eccentric sight. Experience has abundantly shown those forms which tend to progressive blindness to be characterized by an early narrowing of the field of vision, a preponderating loss of sensibility of the peripheric portions of the retina. It is, *à priori*, readily imaginable that, in a progressing atrophy of the nervous elements, those regions should first suffer which are most remote from the nutritive and functional centre, and that thus the extinction of power should be successive and, to a certain extent, centripetal.

Inasmuch, therefore, as extreme significance is to be attached to a defective or diminished peripheric vision, our methods of diagnosticating the same must be as perfect as possible. Ordinary daylight is insufficient to detect slight defects in making a general examination of the periphery of the field of vision. This must rather be conducted in a darkened room where light proceeds from but a single source. Where absolute accuracy is desirable, the "graduated lamp"† may be used; the diaphragm being set at 100, and a black

---

* From this class we of course exclude all those affections which proceed from visible changes in the refractive media, or in the internal structure of the eye; as also cases of neuroretinitis and embolia.

† A lamp so arranged as to diffuse a greater or less amount of light, according to the size of the opening in the movable diaphragm which is placed before it.—TRANSLATOR.

paper without gloss being held before the patient (of course at a fixed distance). The limits of the field of vision are ascertained by means of white balls, set on a black rod and gradually removed from the point of fixation. To ascertain the angle of distinction in eccentric vision, the balls may be placed on the two extremities of a blackened pair of compasses.

The results of this, or a similar examination of the limits of the field of vision may be set down under three heads :—

1. Peripheric vision is relatively the same as in a sound eye.

2. Peripheric vision has suffered a diminution, equal, however, in every direction, and relatively of slighter amount than the derangement of central vision.

3. The derangement of eccentric vision is unequal, i. e., most marked in one direction, or in certain directions; extends from the edge over the surface of the field of vision, and plays no longer a subordinate *rôle* in comparison with the derangement of central vision.

In the first case we designate the limits of the field of vision as *absolutely normal,* in the second as *relatively normal,* in the third as *abnormal.* In considering cases by themselves, we will enter into a more detailed explanation of these distinctions.

Where the limits of the field of vision are absolutely normal (1), progressive atrophy, i. e., actual amaurosis, need never be suspected, provided only that the affection has in other respects properly declared itself. I do not undertake to deny that there is an initial stage where clearly defined symptoms may not yet have become developed, and it is well to avoid giving a decided opinion in a case of recent and imperfectly developed amblyopia. Suppose an instance where, within a few weeks, the acuteness of vision has fallen only to $\frac{2}{3}$, and the limits of the field of vision are absolutely normal, we are still in no condition to give a positive opinion for the possibility of a further diminution of the acuteness of vision, and ultimate encroachments on the periphery of the field of vision is by no means excluded. Has, on the other hand, an affection existed several months, and does the periphery of the field of vision remain absolutely normal, even though the marked diminution of the acuteness of vision to $\frac{1}{6}$, $\frac{1}{10}$, or even less, shows the settled nature of the difficulty, we may safely conclude that the process is not one of progressive atrophy (amaurosis). And thus it may be seen that we are unable to approach a comparatively light case without some de-gree of apprehension, and yet may give a favorable opinion in an apparently graver condition of things. In speaking, however, here of a favorable prognosis, I wish to be understood as referring only to the approach of amaurosis, and not of an entire recovery. We meet a variety of amblyopic conditions, where the periphery of the field of vision is intact, but where central or eccentric scotomas exist, which baffle all treatment and, after reaching a certain height or diminishing within certain bounds, remain permanent. Of this, more hereafter.

If the field of vision is *relatively normal* (2), i. e., if an equal loss has taken place in all portions of the periphery, slight in compari-son with the derangement of central vision, the affair is a dubious one., We may get information from a simultaneous glance at the optic nerve, the age and the manner in which the disease has developed itself, but not from the functional derangements alone. Considered by it-self, this form is by no means a serious one, being the same that occurs in loss of retinal sensibility; for example, in long-continued cases of loss of vision of one eye, and may be artificially produced by bringing a piece of deeply smoked glass before the eye. Just, then, as eyes excluded from the act of common vision by a cause still in force—strabismus, for example—may have eccentric vision outwards predominate over that inwards, and finally over the cen-tral, and may thus become progressively amaurotic, so we see, in the case of amblyopic eyes with a relatively normal field of vision, the cause of the disease still existing (as in the case of drunkards), that progressive atrophy may result. On the other hand, we may witness complete restitution, as in anopsia. In fact, where the field of vision is still relatively normal, we are not dealing with a characteristic case of progressive atrophy. A rational and discriminating system of treatment is here specially called for.

And, finally, if the limits of the field of vision be *abnormal* (3), the case assumes a more threatening aspect; though it would be going much too far to regard all belonging to this class as necessarily tending to a fatal termination. We have to first take into account the manner in which the field of vision is contracted, next to compare this with the acuteness of vision in the centre of the field; thirdly, to regard the appearance of the papilla; fourthly, the manner of development of the disease. The last two points will be considered in another con-nection. The following remarks pertain to the first two.

If the limitation of the field of vision be on one and the same side of the body, for example on the right (or right and downwards, right and upwards), and if central vision is normal, or nearly so, then only one tractus opticus is affected. The disease, as such, may end in complete hemiopia on corresponding sides, never, however, in blindness (Case IV.). An extremely exceptional state of things (and one that admits as yet of no anatomical explanation) is the finding of limitations in the field upwards or downwards, on one or both sides, in the last event symmetrical, which, when they occur separated by a sharply-defined boundary from that portion of the field retaining its normal functions, and when vision is of normal acuteness, give no reason for suspecting progressive atrophy. Actual concentric narrowing of the field of vision sometimes may be observed to result in blindness, the field growing gradually smaller, but preserving its form (as in exceptional cases of glaucoma), although such a state of things seems in general to depend on causes which may be arrested in their operation and even removed. The latter may with some confidence be looked for when the papilla has preserved a normal appearance, when the acuteness of vision has become only moderately impaired, and when by the use of dark blue glasses (shade No. 6 to No. 8) a partial increase of peripheric vision may be effected (as in hysterical anæsthesia of the retina and a peculiar form met with in nervous children). The most unfavorable cases are those where an irregular lateral contraction of the field of vision has taken place, attacking the two eyes either simultaneously or successively, and after such a fashion that the periphery of the field of vision becomes principally impaired inwards or outwards (or in an intermediate direction, outwards and downwards, inwards and upwards, &c.); I say *principally* impaired, for besides the marked contraction on one side there is generally a lack of distinct perception in other directions. The principal distinction between these limitations and those described in hemiopia consists in the fact that the portion whose function has become impaired is never separated by a sharply-defined line of demarcation from a part retaining its normal functions, but that a gradual transition takes place through a tract the activity of which grows greater towards the centre and less towards the periphery. The graver forms of amaurosis generally run such a course as to have seriously affected the first eye, having caused, for example, the contraction of the field of vision to have already passed the point of fixation, when the second eye begins to be involved (see below). In order to be assured of the soundness of the second eye, its field of vision must be closely scanned in that quarter where ominous symptoms are most to be feared. Was the first eye first invaded by a cutting off of the field of vision downwards and inwards, the periphery of that of the second eye must be carefully watched in the same region, and so must the temporal border if the temporal side of the other eye was first affected. These observations are all the more essential because, why we know not, an interval of several or even of many years often elapses between the affection of the two sides, while, on the other hand, one may follow on the heels of the other. It most frequently happens in amaurosis that the first derangement occurs on the nasal border of the field of vision, and that the temporal half holds out the longest. This may be owing to the anatomical arrangement of the optic-nerve expansion and to its physiological consequences, inasmuch as that lateral portion of the field of vision which belongs to each eye for itself is furnished exclusively by the distribution of the fibres on the nasal side. Although, indeed, in processes involving atrophy, so many cases occur where an exactly opposite condition of things is found, viz., the occurrence of the first derangement in the temporal half of the field of vision, that we cannot, as in glaucoma, name them exceptions to a general rule, but must rather look upon them as in the minority. Should it happen that the second eye should not become affected on the same side (medial or temporal), but rather on the opposite, i. e., the side corresponding with that side of the body, the case might, as in the former instance, be one involving only one tractus opticus, developing itself successively in the fasciculus lateralis and cruciatus, and threatening consequently not blindness, but only hemiopia on corresponding sides; in this hope, however, we only dare indulge when, 1st, the limitation of the field of vision on the first eye does not cross the vertical line of separation passing through the point of fixation; 2d, when the acuteness of vision has not materially diminished, perhaps not below $\frac{1}{3}$ or $\frac{1}{4}$. If the case be otherwise, this mode of development, relatively infrequent though it be, must excite the suspicion of approaching atrophy.

Is the survey of the *periphery of the field of vision* of the first importance as regards progressive atrophy, so, on the other hand, may we deduce from a knowledge of its *continuity* very important inferences as regards the question of recovery. It may be generally laid down that those cases of amblyopia in which the continuity of the field is nearest its normal state, offer the best chance of recovery (Case I.). Accordingly if, with a diminution of the acuteness of vision, we find the eccentric acuteness correspondingly affected, so that, in accordance with the common law, it gradually diminishes towards the periphery and does not in certain spots undergo interruptions or sudden diminutions, we are at liberty to form a more favorable prognosis than if central acuteness of vision had been impaired over a defined spot, sharply separated from the surrounding parts (central blur, central scotoma, central defect), or if a slight impairment of acuteness of vision were accompanied by eccentric interruptions. Causes of disease seem often to be at work, in the case of central or eccentric scotoma, which lead to a permanent impairment of sensitiveness. We often witness, at any rate, the persistence of such a state of things, and are therefore in a situation to give a prognosis only favorable as regards blindness, doubtful as regards ultimate recovery (Case III.). In such cases the state of the papilla and the course of the disease afford us more tangible information. I again insist on the fact that, in order to give a favorable prognosis as regards blindness in cases of central and eccentric scotoma, we must have a satisfactory amount of eccentric vision beyond the scotoma, especially towards the periphery of the field. But if the case be one of scotoma, beyond which and in certain directions, even to the periphery of the field, the eccentric vision has become impaired, we have generally to do with a form of progressive atrophy (Case V.). So, too, must we carefully investigate cases of eccentric scotoma, occupying corresponding places in the field of vision, for example, on each side the lower part. If the eccentric vision in the neighborhood of the scotoma, in the last-cited case below, is entirely normal, blindness is not to be feared, indeed the capacity of the layer of fibres corresponding to the scotoma must be normal. If the case be otherwise, such a derangement is not infrequently the precursor of an amaurotic affection.

Beside the diminution of central and eccentric acuteness of vision, many other features of amblyopic affections have had more or less importance attached to them as influencing the prognosis. This was particularly the case before the invention of the ophthalmoscope, when it was impossible to accurately distinguish amblyopia, in the present sense of the word, from affections of the internal structures, from inflammations of the optic nerve, and even in part from impairments in the transparency of the refractive media. Since the classification has been completed, the worthlessness of most of these symptoms has become manifest. Let me briefly refer to some of them. The *subjective appearances of light* both plain and colored, as also in the form of subjective pictures, occur to any extent in comparatively few amblyopic cases, and, when present, give more information as to the accompanying condition of the brain itself (as in encephalitis, delirium tremens) than as to the state of the amblyopia. Vastly more importance is to be attached to these phenomena in diseases of the deep-seated membranes, as they here not only sometimes indicate the duration of the progressive periods, but are even of importance as prodromal symptoms (as, for example, in cases of separation of the retina, where they may occur as white balls, drops, or in crescentic shapes, which I attribute to the tension near the equator). The weight laid on these symptoms formerly in cases of amaurosis is due in great measure to the fact that these cases were confounded with affections of the deeper structures. Again, these appearances may be present in their most intense form in an affection, which (so far as it occurs idiopathically, involving neither the field nor the acuteness of vision) may be of some significance as relating to the cerebral economy, but never passes into amaurosis; I mean the so-called hyperæsthesia retinæ. In cases of genuine amblyopia, misty or smoky vision is most intimately connected with a diminution of the central or eccentric acuteness of vision, and often only the direct indication of this on the part of the retina.

It is at this day hardly necessary to state that the pronounced mouches volantes we often meet in normal eyes (*myodesopsia*) are in no wise indicative of amaurosis. Their cause is partly optical, alterations of the state of refraction, of the range of accommodation, irregularities in the refractive media all facilitating the production of entoptical shadows of diffraction, partly depends on hyperæsthesia of the retina, where shadows ordinarily invisible are brought out by the attention concentrated on them.

The *favorable influence of convex glasses* in amblyopia, as facilitating the recognition and survey of letters, has been used to aid the prognosis. This experiment, leaving out of account any mere correction of refraction, may be used in some cases to demonstrate the continuance of the acuteness of vision, and the full activity of the retina at its centre, gives us, however, in the end, no results that we could not have obtained from the examination of the central and eccentric vision.

The old division of amaurosis into *active* and *passive* has practically passed away; although it is not to be denied that in different forms, the relations of the acuteness and field of vision being apparently the same, the capacity of perception is very differently affected by the amount of illumination, and that in this sense we may be said to be dealing with active and passive forms. It is well known that, in the case of a normal eye, the amount of illumination may vary within tolerably wide bounds; for example, from bright daylight to much less, without exerting a perceptible influence on the acuteness of central or eccentric vision. It first becomes felt (in accordance with a yet undiscovered law) when the amount of illumination is reduced within certain bounds—for example, approaches twilight. In the case of amblyopic eyes we find considerable fluctuations often produced by slight changes in the amount of natural, and generally by greater variations in the amount of artificial light. Where a decided diminu-

12

tion is thus observed, we are justified in speaking of torpor of the retina in connection with the other circumstances of the case. Taking into account the prevalent amount of central and eccentric vision, this "torpor" consists in an abnormally rapid diminution of each (or particularly of eccentric vision), the amount of illumination being diminished. On the other hand, cases are to be met with where perception is not diminished, or is even increased by a withdrawal of illumination that normal eyes would sensibly feel.* In these rare cases we find that in the twilight or while looking through deep-blue glasses distinctness of vision increases, and we may therefore contrast them as active forms with the more strongly-marked passive, and this all the more because, other things being equal, the prognosis is better.

Color-blindness, a not infrequent accompaniment of amblyopic affections, has been recently carefully investigated by Benedict, and especially by Schelske. That it is of any nosological use is, up to the present time, neither possible nor probable, inasmuch as Benedict declares that the pathological color-blindness may give way without any special alteration in the power of perception.

Some, following in the footsteps of Serres d'Uzès, would find a thorough means of diagnosis in the examination of the *phosphenes* produced by pressure. This method is, however, here even less admissible than in cataracta complicata and closure of the pupil, inasmuch as in the unimpeded open determination of the powers of vision we possess a method at once more delicate, better calculated to directly ascertain the true nature of the disease, and more easy to carry into execution. Only in rare cases of peripheric anæsthesia (as, for example, where concentric narrowing of the field of vision occurs in connection with a sudden failure of cutaneous sensibility) are we forced to the interesting conclusion that phosphenes may be produced by pressing against portions of the retina that are insensible to light. In my opinion the solution is to be found in a loss of connection between the layer of rods and the nerve-fibres, and in such a case we may look upon the existence of the phosphene, taken in connection with other circumstances, as rendering the prognosis more favorable.

Finally, an attempt has been made to levy, for diagnostic and prognostic purposes, on the *results of galvanic stimulation*, and Remak has recently published some communications of much interest, the continuation of which we must await, on the different reactions with regard to colors of the centre of the optic nerve in various cases of amblyopia.

We turn now to the second head, namely, the *condition of the optic-nerve entrance*. We are indebted to the ophthalmoscope for having not only set defined bounds to the class of amblyopic affec-

---

* In order to be exact in these statements many details must be added, and above all the intensity (I.) of illumination which is to be placed at the head of the scale of gradually decreasing amounts of light. If a large amount of I. is taken, as much, for example, as would be represented by intense daylight, many amblyopic eyes possessing but confused vision would, as this intensity is decreased, at first seem to lose no distinctness of perception, and only show signs of "torpor" after still further diminution. The remarks in the text above refer to that decrease in central and eccentric vision which follows when we descend the scale from ordinary daylight to that amount of I. which was recommended to be employed in the more delicate examinations of the field of vision.

tions by the exclusion of other intra-ocular diseases, but also for having discovered in the state of the optic papilla diagnostic marks of much importance in separate cases. Our attention is to be paid to four characteristic points, closely connected together and partially dependent the one on the other, viz. (a), *alteration in color ;* (b), *opacity ;* (c), *excavation ;* and (d) *diminution of the calibre of the vessels.*

While the intra-ocular end of the normal optic nerve, let its whole effect be white or yellow, still offers a well-marked reddish tint, which only partially gives way to the more decided bluish-white reflex of the lamina cribrosa, as in the case perhaps of an existing physiological excavation ; in many cases of amblyopia, particularly the more serious, the whole face of the papilla becomes of an intense white. The contrast of color with the adjacent choroid becomes naturally more marked, the edge of the latter appearing to have acquired an increased distinctness. This reflex occurs for a double reason—partly because the atrophy of the papilla lays bare more of the lamina cribrosa, which offers a stronger reflection ; partly because—besides the disappearance of the nerve-fibres—a thickening of the connective-tissue elements takes place in the papillary section of the nerve. These two causes may be concerned together or separately in the change of color. If the first alone acts, the color becomes a bluish-white, and an excavation caused by atrophy results ; if the second acts, the consequence is an intense and pure whiteness of the papilla, the surface of which is smooth ; if both are concerned, the papilla becomes slightly excavated, the lamina cribrosa only visible in detail in places, perhaps through a preëxisting physiological excavation, the remainder being concealed by a layer of white connective-tissue. The reason why sometimes one form and sometimes the other occurs, does not appear to lie in an essential difference in the process of atrophy, but rather, as Schweigger's researches tend to show, in a preëxisting variety in the form of the papilla and in the different physiological intra-ocular pressure, as also in the intermediate effects of a stimulated circulation (on the connective-tissue). The second criterion, the *opacity,* need hardly be referred to after what has been said. The delicate semi-transparency of the normal papilla is naturally wanting when the connective tissue of the lamina cribrosa, or the thickened tissue substituted at the extremity of the nerve, forms the visible boundary. We can no longer, as usual, see the trunks of the vessels dip in,

3

and are involuntarily impressed with the dead look of the opaque substance of the papilla in comparison with its usual appearance of vitality. The third criterion, the *excavation*, has also been sufficiently dwelt upon. As has been said, its presence or absence depends on the more negative or positive condition of the inter-connective tissue in the atrophied extremity of the nerve. A few words, in conclusion, on the *lessened size of the vessels*. It is true that we sometimes find all the vessels, including the main trunks, diminished in calibre, but this is a rule that by no means invariably applies to the latter. We may have a case of complete amaurosis of long standing, in which the optic nerve exhibits every sign of (nervous) atrophy, and where the main vessels have retained their normal diameter. A wholly different state of things obtains when atrophy of the optic nerve is the ultimate result of intra-ocular processes, for example of choroido-retinitis; in such a case the very diminution in size of the main vessels is the most regular and often most striking symptom. The ground of this diversity may be found in the fact that in cases of amaurosis, so far as pathological investigations have yet reached, as also after the division of the optic nerve (Rosow), only the nerve-fibre and ganglion-cell layers atrophy, the rest of the retina maintaining itself; whereas, in cases of choroido-retinitis leading to disappearance of the optic entrance, the reverse is the case, the whole tissue of the retina becoming destroyed. The differences we notice in the calibre of the main vessels in cases of atrophic amaurosis may well (as in the case of the papilla) have their origin in the varying condition of the connective elements of the retina. The diminution in size, if not the complete disappearance of the smaller vascular ramifications on the papilla is, on the other hand, a constant symptom. The normal red shade of the latter depends undoubtedly on the very large number of small vessels interwoven with it, and the absence of these vessels contributes very much to the whiteness found in cases of atrophy. It seems as though these fine branches ramifying on the papilla itself were almost exclusively destined to supply the intra-ocular extremity of the nerve and the fibres expanding in its vicinity, and it would be precisely these ramifications which would be most affected by the disappearance of the fibrous layer; while the largest portion of the bloodvessels of the retina would be preserved for the supply of its other layers. The pallor of the papilla caused by the diminished size of these ramifications has, of course, an appreciable effect on the change of color. If the semi-transparence be tolerably well preserved it may precede other symptoms and be indicative of the first period of the affection, during which an ordinary examination of the field of vision discovers no anomaly; and it needs the more delicate method of investigation to disclose a want of power in the periphery of the field consequent on impairment of the functions of the nerve-fibres.

For the sake of conciseness we will class all those symptoms visible on the papilla, and which may be present in every conceivable degree of the affection, under the head of *atrophic degeneration*, and endeavor to call attention to their practical bearings on the study of amaurosis. And, as has been the case with various ophthalmoscopic symptoms, the multitudinous inferences drawn from which have been greatly cut down by the results of clinical experience, so has it proved with regard to atrophic degeneration of the papilla. Its existence has been used to prove the presence of a process necessarily resulting in blindness, and to a certain extent has been employed as a material foundation for amaurosis. We enter a decided protest against this attempt at identification. It ignores the fact that it is absolutely impossible to tell by looking at an optic nerve whether the atrophic degeneration be progressive or stationary, while here in reality lies the pith of the whole matter. Such conclusions are only allowable when the results of the functional investigation and the manner of development of the disease are taken into consideration. A total absence of atrophic degeneration, consolatory though it be under certain circumstances, is not sufficient to exclude, among others, the very worst fears (Case VI.). On the other hand, the presence of an appreciable amount of atrophic degeneration certainly proves a derangement of nutrition, with which is linked a functional anomaly; but even when present in a marked degree, we are sometimes able to give a prognosis favorable as far as regards ultimate blindness (Cases III. and IV.).

For the sake of explanation, I must return to the impairments of function already spoken of. The coëxistence of atrophic degeneration with a field of vision absolutely normal at its periphery, does not necessitate an unfavorable prognosis as far as regards blindness. The question, then, becomes one of defects of continuity in the field of vision—as, for example, of large central scotomas, a prolonged existence of which exercises a perceptible influence on the appearance of the optic nerve. One of our cases (III.) is an example of this. On the other hand, the presence of atrophic degeneration decidedly complicates the prognosis as regards recovery, and central scotomas thus accompanied are, as a rule, capable of but slight improvement. If the field of vision is relatively normal, the presence of atrophic degeneration weighs heavily in the balance; and this not only because it diminishes the hope of recovery that would otherwise exist, but also because the fear of progressive failure of vision is thus brought much nearer home. In amblyopia potatorum, for example, the appearance of the papilla furnishes an almost entirely reliable test as to the curability or incurability of the case; the same is true in those cases of amblyopia, with a relatively normal field of vision, dependent on chronic meningitis. In connection with a review of the course of the disease, which remains to be treated of, it entirely settles the matter. If the periphery of the field of vision is abnormal, we must make a circumstantial investigation. In hemiopia resulting from the paralysis of one tractus opticus, the presence of a large amount of atrophic degeneration may not alter the prognosis (Case IV.). It would be a grave mistake to deduce from this a necessary progressive change for the worse. Special weight, however, is to be attached to the condition of things in certain cases of concentric contraction of the field of vision, which admit of a tolerably favorable prognosis in the case of delicate children, or individuals of a pronounced nervous temperament, particularly women, provided the optic nerve has remained normal (Case VII.); otherwise, however, may justly give rise to apprehension. The absence of any particular symptoms in fresh cases of amblyopia or amaurosis, with contraction of the field of vision, should not, of course, lead us to indulge in any false feelings of security (Case VI.). Such symptoms need some time (occasionally, however, only a few weeks) for their development, and accordingly not infrequently follow slowly behind the functional failure. On the other hand, if the derangement have existed some time without any implication of the optic nerve, this fact, taken in connection with the other circumstances of the case, may afford us reasonable grounds for encouragement. In fact, a grave form of amaurosis, with an already considerable impairment of the field and acuteness of vision, can hardly have existed several months without having set its seal on the papilla. In a case of advanced impairment of vision of long duration, the papilla is found entirely intact (putting malingering out of the question) only in those rare and curable cases which I denominate proper anæsthesia of the retina. The ascertainment of the condition of things in the second eye is of special importance in those ominous forms which in the strictest sense deserve the name of amaurosis or progressive atrophy. Here a degree of pallor of the papilla, perhaps even a sinking below its proper level, may sometimes be found in that very early stage, when it needs a strict investigation to discover a torpid condition of the retina at the periphery of the field of vision.

Other symptoms about the papilla have had importance attributed to them in cases of amblyopia. Especially was for a long time an infinite deal said of *hyperæmia of the papilla*, as indicative of the first congestive stage of the disease. It is certain that a want of a sufficiently comprehensive survey of the numerous physiological variations has caused much that was normal to be looked upon as morbid. I have no intention here of detailing the ophthalmoscopic symptoms of true hyperæmia of the papilla, and will only state that this may help to explain the reason of amblyopia (as, for example, in intercranial congestion), but will hardly of itself account for a considerable diminution of perception. This refers, of course, to anomalies in the fulness of the vessels without any structural change. Moreover, hyperæmia of the papilla occurs much more frequently as a consequence of undue exertion of the eyes, for example in derangements of the accommodation, than in amblyopia, and is entirely absent in the fatal forms of progressive atrophy.

It has also been thought that *anomalies in the limits of the papilla* might be found in many cases of amblyopia. A loss of transparency in the connective tissue of the superjacent retina, such as often (see above) happens on the papilla itself, would in fact cause a loss of distinctness in the margin of the choroid on account of the increased retinal reflex. It is a very important fact, however, as regards the diagnosis, that this loss of transparency in cases of amblyopia is confined to the papilla, and depends perhaps on a separate nutrition of its connective tissue with implication of those vessels passing over from the choroid (Leber). If atrophy of the nerve depend on an intra-ocular cause, on choroido-retinitis, or particularly on neuro-retinitis, the matter becomes a different one, and it is then the diminished transparency of the adjacent retina, surrounding the papilla with a narrow border or a broader ring (beside changes in the vessels) which enable us to watch this method of development for a long time after the original process has run its course.

Finally, a real *diminution* in the diameter of the papilla has been supposed to take place in cases of atrophy, but neither pathological anatomy nor clinical observation have ascertained anything of importance in this particular.

The third chief factor in aiding us to form an opinion in amblyopic affections is *the mode of development of the disease* and the accompanying symptoms. To enumerate the many points which are here

18

of importance were impossible without going over the entire ground, and we will confine ourselves, at least on this occasion, to a few practical hints.

And, first, we observe cases where the trouble developes itself in the shape of sharply-defined hemiopic or concentric limitations of the field of vision; also of central scotoma and even of total blindness, in either case occurring suddenly or very quickly (i. e., in a few moments, hours or days). It was in pathology formerly the custom to connect any sudden occurrence with a hæmorrhagic effusion, and thus fared these cases. But only those affections where the hemiopia occurred on corresponding sides could be ascribed to apoplexy in the anatomical sense of the word, and even this connection has its exceptions.*

Where we have to deal with a double central scotoma, with sudden blindness of one or both eyes, the hæmorrhagic hypothesis is seldom admissible. We could, indeed, hardly point out its locality without ignoring the information derived up to the present from pathological anatomy. If we have other symptoms, indeed, of a preëxisting or accompanying disease, such as a process at the base of the brain, or the products of encephalitis, we are able to give a somewhat definite opinion as regards the cause of the sudden derangement of vision. But this is by no means generally the case. I have so often seen all these forms occur in entire health, and accompanied by symptoms otherwise so vague, that I must frankly admit my ignorance in general as to their cause. Their connection with hæmatemesis, gastric derangements, or the acute exanthemata, when they thus occur, is still unexplained. An attempt has necessarily been made to take refuge in the favorite hypothesis of a vaso-motory influence and spasmodic contractions in the vessels thereon dependent; an hypothesis to which some accompanying circumstances in the condition of the retina and the general system seem occasionally to give a color. Even the fact of a symmetrical development in each retina, or in the cerebral source of each, might to a certain extent be plausibly explained by nervous influences similarly symmetrical. But when shall we succeed in fructifying this treacherous soil with the seed of sound fact? The significance and the course of these processes is

* We may even see true hemiopia, i. e., want of sensibility of one tractus opticus, as an accompaniment of migraine, of course distinguishing it from that false hemiopia, the seeing halves of objects, which sometimes occurs with migraine.

as uncertain as was their theoretical contemplation. In two cases attacked with apparently identical symptoms, there may result complete recovery, or the affection may persist and atrophy of the optic nerve become developed. The prognosis of the disease must, therefore, depend on its progress. Sudden cases of blindness in each eye seem to me to result more favorably in children than in adults. I shall embrace a favorable opportunity of publishing some facts in this connection. An absolute failure of quantitative perception of light, occurring in a case of sudden blindness, and having lasted one or several weeks, need not necessarily be regarded as hopeless. If it lasts longer, and is accompanied by atrophic degeneration of the optic nerve, the chance of course diminishes. It is favorable in the cases of derangement of vision of sudden occurrence after mental excitement, in which are found remarkable variability in the field of vision, entire retention of the phosphenes, even in the blind portions of the retina, and where absolute darkness exercises a favorable influence—a form of anæsthesia which sometimes accompanies cutaneous insensibility to pain, and which especially invites a solution on a vaso-motory basis. As regards central scotoma of sudden formation, I can give neither a cause nor a prognosis. I have seen it both disappear and remain fixed; never, however, lead to blindness when once its functional characteristics had for some time maintained themselves.

In comparison with the ordinary *more gradual* development, these cases of sudden or rapid occurrence must be looked upon as exceptional. Those cases of amblyopia in which the periphery of the field of vision proves *normal* or *relatively normal*, and the perception throughout the continuity of the field satisfactory, ordinarily develop themselves during several or many months till they either reach a certain point, or, the cause of the disease continuing active, assume other forms. Thus in cases where the periphery of the field of vision is absolutely normal, if the acuteness of vision continues to diminish, a central scotoma—previously impossible to define—becomes gradually developed. Those cases, on the contrary, where the field of vision is relatively normal, may pass into the amaurotic form with atrophic degeneration of the papilla, the field of vision becoming contracted (see above). As a rule, and in contrast with genuine atrophy, the affection in these cases which seem more fa-vorable at the outset, developes itself equally in each eye. As far as regards the chance of ultimate blindness, the fact of the disease having remained a long time at the same point is materially favorable. The more serious forms are, it is true, liable to interruptions, which however, if carefully watched, are seldom found to last more than a few months. The fact that for a long time no change has occurred need not dampen the hope of recovery, provided the optic nerve appears normal, and no contraction of or interruption in the field of vision has taken place. On the contrary, we generally find these forms may be cured by a proper attention to the predominant cause. The immoderate use of alcoholic liquors, frequent indulgence in strong cigars, pelvic obstructions, catamenial derangements, cold extremities, suppression of habitual hæmorrhagic discharges or of pathological and physiological secretions, venereal excesses, irregular sleep and immoderate use of the eyes sometimes exert a separate, sometimes a combined effect, and it is then difficult to assign each their part.* The more, then, taking a general survey of the case, it is possible to ascertain and attend to these causes, the more confidently may we pronounce our judgment. If, partly from the above-named causes, and partly from some of which we are ignorant, symptoms have become developed which render probable the existence of a chronic meningitis, such as violent attacks of recurrent headache, with sensibility of the head to the touch, a sense of confusion, &c., a more cautious prognosis should be given; for although we often succeed in curing such forms of amblyopia by the use of powerful derivative† agents, they sometimes, however, assume a less favorable type. The same applies to amblyopia, although the functional derangement be slight, occurring after severe attacks of sickness—for example, after typhus and erysipelas of the head.

Of particular importance is the process of development in cases of *contraction of the field of vision.* And here should be stated, with reference to hemiopic limitations, that—putting out of the question cases of apoplexy and encephalitis—they may sometimes be induced by idiopathic and even transitory affections of an optic nerve tract. These generally depend on syphilis, in isolated cases, how-

* The fact which cannot be denied, that amblyopia affects men more frequently than women, the proportion with us being about 4 to 1, has seemed to justify the conclusion that smoking is one of the chief causes; many of the other conditions, however, are especially applicable to men, and, in my opinion, excessive smoking must be regarded, in the majority of cases, as simply among the active causes.

† Particularly the seton, which is here especially indicated, not, however, in genuine atrophy. Further, we may use aperients and a course of sublimate.

ever, certain inexplicable nervous influences come into play, and neither the subsequent course of, nor recovery from the disease brings them to light. My experience teaches the development of such favorable cases to be so relatively rapid that the acme may be reached in a few weeks. Similar, but very rare cases of temporal hemiopia, of relatively quick development, are observed, in which—to judge from the entire subsidence of the symptoms—a transitory affection of the fasciculi cruciati must have been the cause. I by no means intend to assert that it is always possible to distinguish soon after its commencement, between this and the more serious affections; but among the more important criteria for determining this as soon as possible, appear to me to be the rapid, as well as nearly simultaneous and symmetrical development in both eyes; the relatively acute central vision, which seldom falls below $\frac{1}{4}$ or $\frac{1}{8}$, and the entire integrity of the papilla after a lapse of several weeks. When within a relatively short time one eye becomes entirely blind, while the other for months afterwards remains intact, the prognosis with regard to the latter is much more favorable than if the disease had progressed insidiously and the first eye perhaps become only partially blind. We have in the first instance to refer the atrophy to one optic nerve proceeding from the chiasma, inasmuch as it would be highly improbable that an affection of the fasciculus lateralis or fasciculus cruciatus alone, and not of the contiguous fibres, should be in question.

If considerable derangements of vision, with contractions of the field of vision of one kind or another, follow on acute and severe cerebral symptoms bearing the type of encephalo-meningitis, we have no right to deduce progressive atrophy from either the unfavorable functional derangements nor from the atrophic degeneration of the papilla. We see, indeed, not infrequently, and in spite of all this, that gradual improvement up to a certain permanent point takes place. I was consulted a week ago by a person 24 years of age, who, at the age of 12, had suffered from a severe affection of the brain, as a direct consequence of which, the left eye had become entirely blind, and the right limited to a small, central field of vision, enabling him with much difficulty to find his way about, and to distinguish the larger letters. The optic papilla of this eye had undergone important changes; the functional state, however, during the

4

last twelve years, had remained the same. We must make it, as a general rule, the object of a careful examination to decide whether the cause of the derangement of vision has passed away or is still active. In the first case we often have to do with a result dependent on destruction of a portion of the conductive elements, and there is hardly a form of amblyopia amaurotica which may not to a certain extent be thus regarded. The same considerations apply to the question of recovery. Complete blindness may supervene on acute cerebral disease, quantitative perception of light may be entirely lost for several weeks, or in exceptional cases, even months, and yet vision be partially restored. My only intention in stating this is to warn against forming a fatal prognosis too quickly. On the other hand, illusory hopes must not be indulged in; if absolute blindness have lasted a considerable time and the papilla become degenerated, the prospect is extremely bad, and at the best becomes confined to the reinstalment of small portions of the field of vision. A marked distinction, however, still exists between processes of this class and those which progress insidiously and without marked cerebral symptoms, inasmuch as these, in no stage of their development, admit of amelioration, and are at the utmost capable of being but temporarily arrested.

The course of the most desperate form of amaurosis is as follows. Slowly, but not regularly so, in the course of months or years, the field of vision of the first eye becomes contracted (generally irregularly, laterally), its acuteness of vision diminishes, atrophic degeneration of the papilla takes place, and the organ is lost, while after the first eye has begun to be affected, sometimes not till after its entire loss, the second commences to run the same course. These cases are, indeed, utterly hopeless; they are regarded as a *noli me tangere* by the experienced physician, who cautiously refrains from active treatment, knowing that this may easily harm, and at the best can be but of little service.

A few remarks may be added on the nature of amaurosis. Where there are no objective intra-ocular symptoms in cases of impaired vision, we are apt to speak of cerebral or spinal amaurosis. In the more favorable and curable forms of amblyopia there must exist anomalies in the circulation or nutrition, to which pathological anatomy gives in general no satisfactory solution. In those cases, too,

23

progressing towards blindness and attended with disorganization
of the papilla, we are by no means invariably able to find evidences of
change in the central organs. Atrophy of the nerve-fibre and ganglion-cell layers of the retina is generally found; at the same time,
also, atrophy of the optic nerves, sometimes stopping at the chiasma,
sometimes prolonged beyond, coupled with atrophic alterations in
the thalami and corpora quadrigemina. If, accordingly, the amaurotic process is to be regarded as no more than a progressive atrophy of the optic nerve and retina, an atrophy the course of which
—whether towards or from the centre—is still doubtful, it is yet
undeniable that this partial atrophy is associated in a considerable
proportion of the cases with a widely diffused affection of the central nervous system, and may therefore be regarded in some cases
as a remote consequence, in others as simply indicating the region
in which the disease first makes itself felt. Especially interesting
is the connection of progressive amaurosis with paralysis and mental alienation, and thus with gray degeneration of the spinal cord.
It is well known that amaurotic affections not infrequently supervene
on disordered states of the intellect; but sufficient stress has not
been laid upon the fact that a large number of those attacked with
amaurosis, who were in perfect possession of their faculties when
first affected, are subsequently the victims of dementia. While, therefore, amaurosis may not infrequently be looked upon as a premonitory symptom of mental derangement, the reverse is almost without
exception the case in gray degeneration of the spinal cord. Characteristic symptoms of the spinal affection (impairment of sensibility)
have become settled before the advent of the amaurotic affection.
This may be explained by the anatomical fact that the course of the
degeneration is from the vertebral column towards the interior of
the skull. Of all the many cases of spinal amaurosis (forming, as
they do, some thirty per cent. of the graver forms of progressive
amaurosis) which came under my observation, I can recollect but
two instances where the disease progressed in an opposite direction.
In the one the amaurotic long preceded the spinal affection, the patient having been entirely blind five years before the first eccentric
pains, followed by the usual signs of tabes, made their appearance.
In the other, the amaurotic affection had also lasted several years,
but at the time of occurrence of the spinal disease there remained
some perception of light in one eye.

As many different opinions prevail concerning the minute anatomy of progressive atrophy of the optic nerve, as with regard to its kindred affection—gray degeneration. Have we to do with a primary inflammation in the interstitial connective tissue of the nerve, terminating in the disappearance of the conductive elements, or are we to regard the disease as a genuine atrophy and to consider the connective tissue, which has taken the place of the nervous elements, simply as a supplementary structure? Certain symptoms during the course of the disease, such as attacks of frontal headache, altered demeanor, sleepiness, which often disappear as the blindness progresses, offer us a strong inducement to clinically maintain the hypothesis of a primary stage of inflammation. Still we must admit that these signs allow of a double interpretation, and that the violent eccentric pains which accompany tabes dorsalis give much more ground for supposing previous inflammation. And yet the fashionable theory of this disease inclines more than ever to genuine atrophy. A decision based on anatomical grounds, which should be final, is a matter of great difficulty, the opportunities of examining the disease in its earlier stages being so rare. It may be regarded as a fixed fact that there is no question of an inflammation of the nerve connective tissue, in the ordinary sense of the word, in amaurosis or tabes dorsalis. The best proof of this is furnished by the inspection of the optic papilla, a thing possible from the first stage of the disease. The changes here are essentially different from those of neuritis, of which we may obtain so perfect a picture. But we see that the change in the connective tissue is not always the same. Sometimes we have a simple loss of substance (atrophic excavation), the most perfect type of a true atrophic process; sometimes there occurs a gradual consolidation of the connective tissue (the papilla growing smooth, white and opaque, while the lamina cribrosa becomes hidden), in which case the connective tissue may be the seat of some very delicate inflammatory process, very different from the ordinary one. But enough of this. We have speculated beyond the limits allowed by our empirically acquired facts, and ventilated certain obscure questions more than in the present state of science may be profitably done. Let us return to the examination of cases, and attempt to utilize what has been brought forward in the study of the various forms of disease.

25

CASE I.

Curative form of Congestive Amblyopia, with Normal Field of Vision.

Florian M., railway employée, æt. 49, of healthy appearance, cheeks and extremity of nose of rather a venous redness, comes on account of impaired vision of both eyes. This has been creeping on for the last twenty months; at first very gradually, during the last four months, however, has perceptibly increased. The functional examination shows that the acuteness of vision (tried by average daylight) has fallen in the right eye to $\frac{1}{6}$, in the left to $\frac{1}{4}$. The periphery of the field of vision (examined by softened lamplight) proves to be *absolutely normal*, and in accordance with this there can be found neither interruptions, misty spots, nor breaks in the acuteness of eccentric vision. Inspection discovers nothing abnormal about the exterior or interior of the eye; indeed, in spite of the already considerable duration of the disease, the papilla retains its delicate red, semi-transparent color.

Its nasal segment is appreciably more reddened than the temporal, owing to the greater number of small vessels and the thicker layer of nerve-fibre in the first direction. To this may be added that the inner is not so sharply defined as the outer edge of the choroid, still a careful inspection reveals the whole contour. Such a condition of the papilla may be regarded as entirely physiological, provided nothing abnormal be found in the vessels and the tissue. It is more or less marked, according to natural varieties in the optic nerve, and shows best generally in a case of physiological excavation. The larger veins are well filled, but are not abnormally tortuous, either as regards their own axis or the plane of the retina. We cannot, then, consider this as a symptom of disease; all the less, in fact, because the patient's complexion bears evidence of habitual venous engorgement.

An investigation of the case shows that the patient has for a long time indulged moderately in brandy, drank large quantities of beer, smoked considerably, and had his rest broken by the duties of his calling. He has suffered from neither digestive nor cerebral disturbance, and enjoyed good health. The result of the examination of the organs of respiration and circulation, the pelvic viscera, the skin and the urine is negative.

As regards the *prognosis* of this case, we may venture on an opinion in every respect favorable. In the first place, a progressive blindness is not to be feared in the least; for in spite of the long continuance of the disease the boundaries of the field of vision prove to be entirely normal. But we may even speak confidently of improvement and restoration of vision, for (1) the continuity of the field of vision is entirely uninterrupted; it is only an affair of general diminution of sensitiveness, no indication of a central scotoma being present; (2) the appearance of the optic papilla is unchanged; notwithstanding the difficulty has lasted two years, there is no trace of atrophic degeneration; (3) in the habits of life of the patient we find palpable causes, amenable to therapeutic influences, and of a kind that experience teaches us may produce derangements capable, to a certain extent, of removal.

It would be a difficult task to accurately explain the *nature* of this affection. No signs of active cerebral congestion exist. The venous redness of the face and the predisposing causes render probable a so-called passive congestion; this expression, however, conveys but a limited amount of information. Suppose the case to be one of an inundation of the central nervous structures with venous blood, or of a want of rapidity in the movement and change of the blood itself, or let there be a diminished activity of function on account of the blood being overloaded with alcoholic and narcotic substances—every one of these suppositions would be explained on the ground of " passive congestion of the brain." The only general and satisfactory use of this term is, therefore, in cases where, there being no signs of active congestion, the functional as well as the nutritive activity of the cerebral centre of the optic nerve has been reduced by influences springing from the source alluded to and acting through the circulation.

How is the *cure* of the case to be effected? First, of course, by paying a proper regard to its cause. The use of alcoholic beverages must be given up, that of tobacco reduced to a minimum, regularity in diet and sleep insisted on. Not infrequently is this course alone sufficient to cause a gradual retrogression of the symptoms in cases of this form of amblyopia. Experience teaches us, however, that there are more efficient means of securing and hastening a favorable result. As regards local depletion, a rapid evacuation of the blood is here of special importance. I was some time since led by this to appreciate the application of the leech of Heurteloup in the treatment of congestive amblyopia, and the method based on this has found much favor with the profession.

A good deal depends here on the manner of application. If the operator to whom we commit the task is not competent to fill a cylinder in a few minutes, the particular advantage* of this method is lost. The application must be made in the evening, so as to give the full benefit of the night's rest. It is, moreover, desirable in many cases that the patient should preserve perfect quiet the next day and remain in a dark room. As the importance of this precaution in such case depends on a careful consideration of the individual circumstances, it is best to regard it as a general rule. It is of consequence on account of the excitability of the cerebral circulation, or—if preferred—of the vaso-motory nerves. In cases where this is considerable, each application is followed by a period of excitement, characterized by derangements of sensibility of every kind, sometimes by subjective appearances of light, and even by some diminution of the acuteness of vision. This period of "reaction," lasting only in exceptional cases more than a day, should be prepared for by entire bodily rest and a strict seclusion from light. This state, moreover, is least marked in cases of amblyopia arising from passive cerebral congestion, and most so in certain affections of the choroid; the period of rest, therefore, is less indispensable in the former case than in the latter. From two to four ounces of blood should be taken at each application. And this should be repeated at intervals of four, six or eight days, according to the constitution of the patient and the duration of the period of reaction. A careful examination of the acuteness of vision instituted directly prior to and two days after the application (the period of reaction being thus passed) will decide the propriety of repetition. Where the acuteness of vision has not been affected by two, at the most by three applications, they should be omitted. But even where a perceptible effect is noticed, I do not advise a too frequent repetition in cases of passive cerebral congestion, although of course the state of the constitution must influence the decision in individual cases. For it has been ascertained that those cases which evince improvement after the first three or four bloodlettings, do better under diaphoretics than by a continuance of the abstraction of blood, and it is our duty—other things being equal—to give the first named the preference as a less severe course.†

* Be it said, in this connection, that the Heurteloup leech, rendering, as it does, most valuable service in congestive amblyopia and chronic affections of the choroid, is by no means as good as the natural one in the treatment of the different forms of ophthalmia. It is here that we derive more advantage from a prolonged suction and a continuous flow of blood than from the rapidity with which it is lost.

† Although warmly recommending the Heurteloup leech in cases of amblyopia with passive cerebral congestion, I am willing to admit that some cases derive more benefit from other measures. This is especially the fact when the excitability of the circulation is extremely pronounced. Here the period of reaction is seen to be abnormally lengthened and indisposed to yield to the desired remission. In such cases a decidedly better result is often obtained from cupping in the neck, in hæmorrhoidal affections from leeches to the anus, and in disorders of menstruation from cupping on the inside of the thighs; indeed, under these circumstances the Heurteloup leech may even increase the disturbance and harm the case. The question so often raised in practice as to whether the removal of blood shall take place from the immediate vicinity or at some distance from the affected part, depends in a measure on the method employed, and especially on the degree of excitability, and considerable caution must be used in answering the question in the first sense. A common mistake

Our diaphoretic treatment was principally (in imitation of many older practitioners) carried out by means of the decoction of Zittmann, not because any specific effect is to be attributed to the separate ingredients of this complex draught, but because experience has shown it to be an excellent diaphoretic of proved worth. The strong decoction, warmed, was therefore given at an early hour to the patient in bed, and its action on the skin assisted by means of woollen coverings and afterwards by elder-tea, while on the other hand no special restrictions with regard to diet, as is generally the custom when this preparation is used, were imposed on the patient; a walk, too, in the afternoon was allowed when the weather was favorable. Of late we have been more sparing in the employment of the decoction of Zittmann, having learned to recognize a thoroughly invaluable therapeutic agent in well-appointed Roman baths.*

The patient was again presented at the clinique four weeks after the commencement of his treatment. His course of life had been regulated and the Heurteloup leech applied three times, in consequence of which his acuteness of vision had successively increased from $\frac{1}{6}$ on one side and $\frac{1}{7}$ on the other to $\frac{1}{4}$ on each. Following this, a Roman bath had been taken every three days. No result followed the first bath, except a headache which lasted twelve hours, owing to the fact that he had not remained long enough in the sweating room and had been disturbed by the application of a cold douche at the wrong time. The next baths, in the administration of which these points were attended to, produced an excellent result, the acuteness of vision having risen, after the fifth had been taken, on

---

in cases of ophthalmia is to place the leeches nearer the eye than consorts with the sensitiveness of this organ and the surrounding structures. Further individual circumstances may sometimes induce us to employ other applications in preference to the Heurteloup leech on the temple, as when the disease seems, for example, to be connected with the cessation of habitual epistaxis, or of a hæmorrhoidal flow, &c.

* My first inducement to employ Roman baths in cases of amblyopia with passive cerebral congestion was furnished me by a patient who had been but moderately benefited by a course of decoction of Zittmann, and who entirely cured himself of his amblyopia by remaining in the warm boiling-room of a sugar-factory, the heat being nearly 40° (122° Fahrenheit). Great advantages as the Roman baths possess in most of these cases over the Russian baths, they have, of course, their contra-indications, among them the more active states of congestion, and especially cardiac and renal affections, an apoplectic tendency and undue excitability of the circulation.

[These "Roman baths" are known in this country as Turkish baths, and differ from the Russian in the employment of dry heat instead of vapor.—TRANSLATOR.]

the right to $\frac{2}{3}$, on the left to more than $\frac{1}{2}$. A speedy and entire recovery is therefore no longer doubtful.

Von Graefe remarks, in conclusion, as follows. Recovery does
not follow as rapidly in all these cases of amblyopia as in the present. If, however, appropriate treatment succeeds in once causing
the affection to take such a turn that the acuteness of vision begins
to increase, we may reckon with nearly entire confidence on a gradual and almost spontaneous improvement, provided the cause of
the difficulty continues to be avoided. The attempt must not, therefore, be made to perfect the result by continuing the employment of
vigorous measures during a prescribed time, but rather, after the first
blow has been administered to the disease, it should be left to itself
awhile, after which the former treatment may be resumed. Thus
the entire amount of previous vision may often be seen to return in
the space of several months or more. On the other hand, a strong
tendency to relapse may be manifested, when, owing to a faltering
resolution or the pressure of circumstances, the patients again come
within the influence of the previous exciting cause of the disease; in
fact, such relapses may exhibit a paralytic tendency, differing from
the previously mild form (as in the case of amblyopia potatorum).
This teaches us how necessary it is to lay particular stress on the
manner of life of such patients and regulate their labor. Finally,
be it remarked that while we very frequently employ the course that
has here proved itself efficacious, it should by no means be regarded
as the exclusive method of treatment in such forms of the affection.
No cases need more individual study than those of amblyopia, and
the physician who after a single and hasty survey undertakes to
give such patients advice available for some time to come, commits
a serious fault. The derangements in the circulation which are here
concerned may spring from the most varied sources, and it is both
perilous and narrow-minded to deduce them by preference from
some particular organic affection, from the liver, for example, or the
alimentary canal, or some irregular course of life. Abdominal disorders, it is true, often play a principal part among the causes, and
in accordance with this view, we often see good effects resulting
from the use of mineral springs, such as Marienbad, Kissingen, Hom-
burg, and in suitable cases particularly Carlsbad; as a general
thing, however—thanks to the prevailing tendency—the functions of
the abdominal viscera occupy too exclusively the attention of the

physician. The weighty functions of the skin and kidneys, so important in their effect on amblyopic affections, are among those seriously neglected through this preference. While the regularity and amount of the fæcal evacuations are anxiously dwelt upon, these principal regulators of the circulation are hardly regarded, as being of minor importance. Even the ways of life, from which but too often proceed the accumulated causes of derangement of the circulation, are often but carelessly scanned by the physician, and as a matter of course made light of by the patient. If the sleep be not particularly disturbed, little is said about it, and yet good and regular sleep affords the most grateful refreshment to the unceasingly active nerve of vision. To be here a successful physician, one must examine with extreme care and weigh the result with foresight and impartiality.

Case II.

Progressive Amaurosis, depending on Atrophy of the Optic Nerves.

Julius M., a sailor, æt. 24, of tolerably robust appearance and a healthy complexion, presents himself on account of a considerable impairment of vision, rendering it already difficult for him to find his way about. According to his own account, the left eye became affected six months, the right four months ago, and in both the rate of progression has been tolerably equal. At present the acuteness of central vision amounts to about $\frac{1}{100}$ on the left side, $\frac{1}{30}$ on the right; it should be stated, however, that in the left eye the acuteness of vision in the direction upwards and outwards is not only relatively but absolutely greater than in that of fixation, so that while fingers are with difficulty counted at a distance of nine inches when held directly before the patient, they are made out in two feet if held at an angle of 20° from the point of fixation. This agrees entirely with the following state of the field of vision. In the left eye the entire inner half is wanting, and a hand held up in good daylight is nowhere visible on the farther side of the vertical line of equal division. In the outer and lower quadrant, too, eccentric vision is extremely imperfect, and becomes entirely defective by reduced lamplight. It is only relatively good in the outer and upper quadrant, and is here in one direction more pronounced than in central vision, as is seen in the tendency to eccentric fixation. A better functional state is found in the right eye. The loss in the field

of vision, proceeding here, too, from the inner edge, does not extend
to the vertical line of equal division, but remains in the plane of
vision about 15° from the point of fixation. Below the plane of
vision, indeed, it approaches nearer this vertical line, while above it
recedes from it. A considerable failure of distinctness in the eccentric vision extends, however, far beyond these limits, it being impossible to count fingers in the immediate vicinity of the nasal side of
the point of fixation. In accordance with this, the defective region
nearly grazes the point of fixation, by diminished lamplight. In the
outer half of the field the eccentric acuteness of vision is relatively,
although here not absolutely, better than the central. The considerable loss of perception, and especially the contraction of the field
of vision, is manifested in the unsteady, tentative gait of the patient.
As the light wanes, he becomes entirely helpless, owing to the above-
mentioned torpidity of the retina, existing, as it does, within a
large portion of the already contracted field of vision.

With the exception of the sluggish pupillary reaction, especially
marked on the left side, nothing abnormal is externally visible. The
ophthalmoscope reveals a normal state of the refractive media and
internal membranes, and in each eye a high degree of atrophic
degeneration of the papilla, under the form of atrophic excavation. The elements of the lamina cribrosa are plainly visible
in the larger and excavated portion of the papilla, on the temporal
side up to the edge. The remaining substance of the papilla, namely, inwards from the vessels, is of an opaque whiteness; the smaller
vessels are wanting; those of medium size are somewhat, the larger,
on the other hand, hardly at all contracted.

An examination of the case shows that, at the commencement of
the change in his vision, the patient suffered from slight frontal headache, increased by stooping and coupled with a sense of confusion;
that for several months, however, these, at the best but slightly pronounced symptoms, have entirely disappeared. At present no trace
of disease is to be found in either any organ or in the physical or
mental functions. He labors, naturally, at present under some mental depression; no more, however, than might reasonably be attributed to the daily failure in his vision. Nothing of consequence
can be ascertained from his previous habits of life, except a not im-
moderately excessive use of tobacco, which, however, he gave up at
the first commencement of the trouble.

The prognosis of this case is altogether a gloomy one. The contraction of the field of vision has taken place in that ominous manner which characterizes *progressive atrophy :* slowly advancing limitation from the inner edge, first on the left side, then symmetrically on the right, besides enormous loss of sight, which has caused on the left side an absolute, and on the right a relative preponderance of eccentric over central vision.

Before, however, giving up all hope, it is our duty to think over all the possible relatively favorable chances, if necessary to exclude them. May we not, perhaps, have to do with one of those cases (already referred to) of contraction of the field of vision resulting from hemiopia, and in which recovery is possible? Unfortunately, our reply must be in the negative; for in the first place the affection has extended on the left side far beyond the vertical line of equal division; in the second, the acuteness of vision at the point of fixation has become, even on the right side, too much affected; in the third, the defective portion of the field is not sharply set apart from the parts that retain their normal function, but is bounded by extensive portions, the acuteness of vision of which is reduced and the torpor typical; fourthly, the development, although more rapid than in the average of cases of progressive atrophy, cannot properly be called acute, i. e., reaching its height in a few weeks; fifthly, we have to do with limitations of the inner portion of the field on each side, whereas such curable or stationary cases have thus far been observed to take the form of hemiopia either temporal or occurring on corresponding sides (very seldom that of defective portions extending upwards or downwards). May not this, we ask again, be perhaps the result of an action already terminated or capable of being arrested, and may it not on this theory be possible to keep what vision exists at present or save an appreciable portion? Neither the course of the disease nor the attendant symptoms support this view. In the case of rapidly-developed amaurotic disorders, accompanied by well-marked cerebral symptoms, the subsidence of which leaves the affection stationary, we have a more favorable foundation to build upon; such is, however, not here the case, the vague indications of cerebral disturbance which previously existed have passed away without making the least impression on the continued development of the amaurosis. In this point of view, its steady increase is discouraging. The less the connection that can be made out with

ulterior cerebral symptoms that possibly admit of relief, the more
the atrophy of the optic nerve plays the part, so to speak, of an independent disease, the more desperate is the prognosis. It is, however, made less certain by the discovery of some particular habit or
way of life of the patient that is manifestly productive of mischief,
if we have reason to trust that the removal of this may affect the
progress of the disease. But even under these circumstances let no
illusory hopes be indulged in. Has the amaurotic affection passed
into the stage of advanced contraction of the field of vision, and
has besides marked degeneration of the papilla taken place, the secondary disease may be known to have attained a fatal indepen-
dence of its cause. But no fact favorable to our case can be made
out in this connection. The only suspicious habit, the undue use of
tobacco, had been relinquished at the commencement of the disease.

Even in the worst cases of progressive atrophy, it may exception-
ally happen that an unexpected pause occurs, after the acuteness of
vision has appreciably diminished. I saw, for example, within a few
days a patient who had been treated by me for spinal amaurosis
eight years before. Vision had at that time become extinct in one
eye, and within a year the other had lost so much as to be only able
to follow the movements of a hand held on the temporal side. By
referring to the notes I took at the time, I have recently ascertained
that this state of the vision, poor as it was, had remained unaltered
during the space of eight years. Such decided pauses are, however,
most exceptional, and I have hardly observed them except in cases
where vision had been reduced to its lowest ebb, where, indeed, in
the popular sense, blindness might be said to have already commenced. They are to be distinguished from the temporary pauses which
last some weeks, rarely from four to six months. These latter happen very frequently, and in the most varied forms of progressive
amblyopia, especially those dependent on spinal disease. But in
our case the advanced atrophic degeneration of the papilla, with the
continuous loss in the field of vision and the failure of all other
symptoms, offer no foundation for these hopes. And although we
are withheld by the variable and obscure nature of these affections
from expressing with too absolute certainty an opinion unfavorable
to the possible preservation of a very slight amount of vision, perhaps quantitative perception of light, it is still most probable that
within a few months, perhaps a little longer, the patient will become
the prey of absolute blindness.

As regards the nature of the disease, it is, in fact, impossible for us to make any other diagnosis than that of *progressive atrophy of the optic nerves.* At present, no symptoms of any other morbid change are to be found. The paroxysms of headache the patient had at the commencement of the disease, and which in an entirely analogous manner are met with in many amaurotic conditions, do not in my opinion afford special indications of an existing irritation in the substance of the brain or in its membranes. I am inclined to explain many such headaches on the ground of the disordered functions of the eyes themselves. When patients begin to lose their vision, and yet concentrate all their faculties on the appreciation of their visual impressions in order to follow their avocations or guide themselves about, there result derangements of sensibility similar to those occurring in diplopia, seeing in circles of dispersion, &c. In such cases we observe that the headache disappears as soon as the patient intermits his efforts to see. If, however, in the face of this, the atrophy or its cause progresses, we have no longer a right to directly connect the headache with the cause alluded to. This, of course, applies only to certain forms of headache, and the true state of the matter must be decided by a review of the whole case, especially by trying the effect of entire darkness. In the present instance we had nothing on which to base our researches, inasmuch as the headache had already disappeared.

That a headache, caused in the first instance by attempts to see, should be increased by stooping, or anything inducing congestion, is of course not surprising. I am willing, too, to grant that the headache itself may be regarded as congestive, so far as the efforts at vision are propagated along the vaso-motory nerves, as may be best seen in the conjunctival vessels. I merely wish to call attention to this point, that the accompanying headache is not always connected with the cause of the amaurotic affection, but may directly depend on the derangement of vision. In other cases of atrophy paroxysms of pain occur, unconnected with the act of vision and evidently attributable to the cause of the malady. Here, too, it remains an open question as to its original inflammatory nature, inasmuch as it seems that the source of such paroxysms may be found as well in genuine atrophy of isolated segments of the brain as in the so nearly-allied gray degeneration of the spinal cord. Finally, I am not disposed to deny that in cases of amaurosis complicated with chronic meningitis, there may occur pains characterized as inflammatory by the sensibility of the head to the touch, a strongly-marked confusion of the faculties—also by the duration and course of the attacks; but I do not regard such a state of things as either regular or frequent in progressive atrophy of the optic nerves.

In the case of this patient there exist at present no symptoms of paralysis of body or mind, and this fact renders it our duty not to venture beyond the diagnosis of atrophy of the optic nerve. But will it be so in the future? It is exceedingly possible that in the course of years mental alienation may become developed, or some further affection of the general system of a paralytic nature. Though, however, this sequence is not infrequently observed, it is still the fact that more than half the cases of amaurosis do not advance beyond the narrow limits of the special affection, and that when death occurs after a lapse of years, an examination reveals either atrophy of the optic nerves or partial atrophy of the portions of the brain connected with them.

As to anything additional on the score of treatment, it is simply necessary to state that all powerful derivative agents, cathartics, setons, mercurials, diaphoretics and depletives, beneficial as they may be in cases of congestive amblyopia (Case I.), are here decidedly injurious. The more we distinguish the different forms the more will this conviction force itself upon us. Everything that suddenly depresses the strength or excites the circulatory system is to be most carefully avoided. We must, however, cease to act on these principles when an investigation of the case causes us to arrive at opposite conclusions, by demonstrating, for example, the existence of chronic meningitis or the suppression of habitual secretions; but even then (considering the state of paralysis that exists) care must be taken that the remedies employed, such as leeches behind the ears, setons and sublimate, be so administered as not to cause sudden depression of the powers. In the average of cases of progressive atrophy, the best means of retarding its progress consists in a mild tonic course—small doses of iron, salt and tonic baths, milk and whey diet; in other words, a nutritious but not stimulating diet, good air, a moderate course of cold bathing and a carefully-regulated amount of light.

In general, a case like the present may be regarded as a *noli me tangere*. We have a right to look upon it with the same dread with which the physicians of other days approached an amaurosis. An inevitable experience it is, indeed, to the ophthalmic surgeon to find such patients leaving their homes and undertaking long journeys in the hope of finding succor; returning, as they certainly must, having accomplished nothing, and often much blinder than before. Such an event has a very depressing effect on the spirits of the patients, not only because their hopes have failed of fruition, but also because gradual loss of vision is relatively more endurable under the circum-stances and exposed to the influences of every-day life.

Central Scotoma, with partial Atrophy of the Optic Nerve, admit-ting only of gradual and partial Improvement.

Alexander K., aged 20, a coachman, and of tolerably robust ap-pearance, comes to us on account of impaired vision of the left eye, coming on, as he states, five months ago, and taking but a few weeks to attain its present development. The right eye, too, sees but im-perfectly; this, however, he says has always been the case.

The functional examination reveals, first, an entirely normal con-dition of the boundaries of each field of vision; the acuteness of vision, however, is reduced in the left eye to $\frac{1}{30}$, in the right to about $\frac{1}{20}$, owing to the presence of central scotomas, which subtend an arc of from $8°$ to $10°$ in the centre of the field. By ordinary light the patient finds it difficult to define these scotomas; by a moderate amount of artificial light, however, he is abundantly able, and it be-comes manifest that the amount of vision just referred to is eccen-tric, inasmuch as within the limits of the central scotoma the patient has but a quantitative perception of light. In making the ophthal-moscopic examination, it is found that a small image of the flame, reflected by means of a plane mirror on the region of the fovea cen-tralis, awakens but a feeble impression. The scotomas are surround-ed by a ring-shaped zone where vision is defective, broader on the inner than on the temporal side.

We hence arrive at the conclusion that the affection of the right eye is by no means congenital, but a matter of recent development, probably simultaneous with that of the left. It is possible that the patient did not notice its coming on, from the fact of this eye being excluded from the act of common vision through a slight divergent strabismus. For amblyopia resulting from exclusion never assumes the shape of central scotoma. In the lighter forms it is character-ized by an equal loss of sensitiveness (diminution of central and eccentric vision, especially the former); in the more serious cases, those occurring with persistent strabismus, by a predominance of the inner over the outer portion of the retina.

The ophthalmoscope reveals an entirely normal state of the re-fractive media and membranes, a physiological excavation (extending here but a short distance outwards from the point of emergence of the vessels); over and above this, however, an undeniably whitish coloration of the papilla, caused by disappearance of the smaller vessels and a small loss of transparency in the tissue, a slight degree, therefore, of atrophic degeneration. Nothing was obtained from a general physical examination. Shortly after this difficulty commenced, the patient had attacks of dizziness and headache, recurring sometimes and lasting even several days; entirely disappearing, however, during the last few months.

An entirely favorable *prognosis,* as far as regards the danger of ultimate blindness, may here be given. As has already been stated, central scotoma, coupled with a field of vision the limits of which are entirely normal, never indicates progressive atrophy. If the disease is in its incipient stage, it is well not to lay too much stress on this fact; not so, however, in a case like the present, where the form is well marked and the affection has culminated. Deterioration, too, is not to be feared. Central scotomas either occur very suddenly, or else they become developed in a few weeks simultaneously or successively in both eyes, increasing outwards from their own centres (not always proceeding straight from the point of fixation); or finally, cases of amblyopia of more prolonged duration, and in which no interruption of the field of vision had been observed, develope at a later period signs of central scotoma after such a manner that while, as a rule, the existing central acuteness of vision is retained, the eccentric improves up to a certain point. We may assume, in all these cases of central scotoma, that where the same state of things has persisted for several months a change for the worse is improbable. The undeniable change in color of the papilla, which is yet to be discussed, by no means necessarily denotes a danger of progressive atrophy (see above). Is the prognosis, then, favorable as regards a worse condition of things, or blindness, the same is not the case as relates to recovery. If central scotomas have lasted beyond a few weeks, and visible degeneration of the papilla has taken place, an expectation of complete recovery is no longer to be indulged in. In general, a slow improvement takes place, sometimes hardly perceptible and always imperfect, the blank caused by the scotoma becoming smaller, and the surrounding ring-shaped zone, where vision was imperfect, gaining in power, while practice gives eccentric vision a more than normal amount of acuteness. It

6

sometimes happens, too, that in the middle of the central scotoma itself a patch clears away, thus allowing of a satisfactory amount of central vision. (This last event, however, occurs much less frequently than in certain cases of central scotoma resulting from circumscribed choroido-retinitis, well known to all.) Particularly important in making up the prognosis is a knowledge of the fact whether vision is relatively most acute centrally, within the scotoma itself, or in an adjoining region. In the case of a scotoma, the angle of aperture of which is 20°, and where vision of $\frac{1}{10}$ corresponds with the point of fixation—the patient "seeing through the spot"—the prognosis as regards recovery is much better than where a scotoma with a small angle of aperture coëxists with vision $\frac{1}{10}$ which is eccentric, i. e., is situated in a region adjoining the scotoma. It is easy to understand that a disease interrupting the transmission of impressions must be much more grave when it brings about a loss of central vision. It has already been stated that this is the condition of things in the present case, and this circumstance, in connection with the fact that the disease has for some time remained at nearly the same level, and degeneration of the papilla has taken place, leads us to infer that at the utmost there may result an almost imperceptible improvement in the first sense alluded to. Whether our patient will, in the course of time, be again in a condition to read ordinary type is a matter of uncertainty.

It is particularly difficult here to enter more minutely into the question of diagnosis. Pathological anatomy has as yet furnished us with no data in such cases of central scotoma, and clinical observation gives us but little theoretical instruction. The fact that both eyes are affected, while other symptoms of central disturbance are wanting, has caused preference to be given to the theory of the existence of some morbid process in the chiasma nervorum opticorum; such an hypothesis, however, I think rests here on slight grounds. Were some material cause, such as apoplexy or impairment of substance, located in the chiasma, the symmetrical character of the affection would seem to me entirely incomprehensible. It appears to me much more probable that a defined cause of the disease holds sway at the cerebral extremity of the optic nerves, and that these regions feel the influence of that symmetrical tendency that governs twin organs of sense, and is often so strikingly shown in the external portions of the eye. But to what particular derangement can we bring the affair home? The instantaneous occurrence of scotomas has caused them to be attributed to hæmorrhagic causes. Against this may be set the limited extent of the affection, as well as its more gradual development in otherwise analogous cases. Moreover, this state of things is never found in connection with other hæmorrhagic diseases of the brain, and is ascertained to relatively occur with most frequency in young patients, with whom the principal predisposing causes of cerebral hæmorrhage are absent. It is not impossible that we have to do with anomalies of another kind in the circulation, with a stimulation affecting certain points of the vasomotory nervous chain, or, in general, with a functional (molecular?) interruption of the transmission of impressions, the first approach of which we are unable to trace to any palpable material cause. The subsequent degeneration of the papilla is consistent with this theory, inasmuch as it is to be regarded as probably consecutive. If we suppose the disease to have proceeded from causes of a vasomotory nature, its symmetrical occurrence at the inner end of each optic nerve would admit of ready explanation. It would thus, too, be rendered comprehensible why central scotomas should sometimes form after an exhausting disease of the entire system, with other signs of vaso-motory disturbance; or after mental affections, with absence of cutaneous sensibility.

The foundation of our diagnosis being thus uncertain, our choice of remedies must be based principally on the state of the constitution, as also on the causes and accompaniments of the disease. If the affection is recent, or if there are symptoms of cerebral congestion, it would be well to try the effect of a few local bloodlettings, by applying leeches behind the ears or the Heurteloup to the temple. Where hæmorrhoidal or catamenial complications are suspected, they of course deserve consideration. If symptoms of mental derangement manifest themselves in connection with the disease, or if cutaneous sensibility becomes impaired, a course of zinc or nitrate of silver should be prescribed. If the skin fulfils its functions irregularly, the effect of a powerful diaphoretic—a Roman bath, for example—should be tried, if no contra-indications exist. If, on the other hand, central scotoma has been developed in the course of a prostrating disease, a tonic course is indicated—small doses of iron, warm baths, nourishing food, residence in the country, &c. We cannot boast of having found any special course of treatment indicated in the case of our patient. His previous attacks of giddiness and headache admit the supposition of congestive antecedents, although susceptible of other explanations.

The patient was shown again two months later. Several local abstractions of blood had meanwhile been made, and a diaphoretic course carried out. The improvement noticed consisted in an increased energy of the indistinct region surrounding the scotoma, without any particular change in the scotoma itself. In consequence of this the acuteness of vision (still eccentric) had increased to $\frac{1}{20}$ on one side and $\frac{1}{16}$ on the other. The patient was now to commence systematic practice of the eyes. For such cases prove that the standard of eccentric vision is set by the amount of its employment, and that the capacity of that portion of the retina adjacent to its centre is capable of being developed beyond the normal bounds. Without ever attaining the normal acuteness of the retinal centre, eccentric vision may double and treble itself in cases of central interruption. This development may be partially instinctive, brought about by the use of the eyes, and partially promoted by systematic practice. We give patients, the acuteness of whose vision is insufficient, ordinary type to read, and then generally magnifying glasses of unusual construction, and based, I think, on a sound principle. Two convex glasses—in this case $+\frac{1}{4}$ and $+\frac{1}{8}$ were selected—are placed an inch apart in a short metallic tube, and the whole furnished with an appropriate handle. We thus secure a considerable amount of magnifying power in connection with less spherical aberration, and a removal to a better distance from the eye than in the case of ordinary convex glasses. I need not add that this system of practice should be restricted at first to a very short time, perhaps two minutes, and to a size of type not barely recognizable, but seen with some degree of readiness.

CASE IV.

Cerebral Hemiopia, occurring on similar Sides, stationary, resulting from an Apoplectic Attack.

August K., weaver of fine cloth, 68 years old, comes to us on account of deranged sight, consisting partly in double vision, partly in a diminution of the power of distinct perception. An examination shows the double vision to depend on a paralysis of the right abducens. The power of motion of the right eye outwards is reduced

2‴ in comparison with that of the left, the patient having, accordingly, homonymous double images, the distance between which increases towards the right. The remaining derangement of vision consists in a slight diminution of its acuteness in each eye to ⅓, and in an entirely symmetrical anomaly of the field of vision. This is considerably contracted in each eye on the left side; eccentric vision is, moreover, indistinct over the whole left half, well up to the vertical line of equal division. Inasmuch as the ophthalmoscopic examination reveals no morbid change, except a partial atrophy of the optic nerve (to be referred to hereafter), this hemiopia of the left side must consequently be referred to a paralysis of the right tractus opticus, and the diagnosis of the whole disease may, therefore, be set forth as a paralysis of the right abducens and a paralysis of the right tractus opticus. A general examination reveals considerable and extensive arterio-sclerosis, hypertrophy of the left ventricle and insufficiency of the aortic valves.

From an analysis of the case we gather the following facts. Rather more than three years ago, the patient had an apoplectic attack, which resulted in hemiplegia as also hemiopia on the left side. He was under our treatment at that time, and the records show that the hemiopia, which at first had been nearly complete (failure of the field of vision up to the vertical line of equal division), gave way during the convalescence to the present point, the acuteness of vision rising from ⅙ to ⅓. Consequently, since that time, his power of sight has remained entirely the same. Fourteen days ago, the patient was compelled to go out in great haste with uncovered head in the midst of a snow storm, got into a profuse perspiration, the next morning had an acute, though not very severe headache, and, while attacked by no proper cerebral symptoms, noticed diplopia, which afterwards became more marked.

The question first arises as to whether any proper connection is to be established between the two paralytic affections, the loss of power in the right tractus opticus and the right abducens. We are of the opinion that this must be answered in the negative, for the following reasons:—

(1.) The hemiopia is evidently to be traced to the results of apoplexy in the right hemisphere. Its sudden approach, attended by cerebral symptoms and succeeded by this particular affection of vision and hemiplegia of the left side, admits of no other inference.

From such an apoplectic source on the right side the paralysis of the right abducens could not evidently proceed, but must be referred to a second effusion on the left side, if connected with any central apoplectic source.

(2.) Supposing any connection to exist between the two affections, as, for example, that a basilar process supervened on the preëxisting cerebral disease of the right side, it would seem remarkable that the sharply-defined traces of the previous attack underwent absolutely no variation, but remained entirely as before. It may be added in this connection that the slight lameness of the left foot, a relic of that attack, has not experienced the slightest modification from the late event.

(3.) It is within the bounds of belief that a circumscribed apoplectic effusion *intra-cerebrum* should give rise to no other symptoms of paralysis than a loss of power in the abducens of the opposite side, as has been seen in cases of facial paralysis; still this isolated action, with no concurrent cerebral symptoms, is in any case improbable. The paralysis of the abducens itself, although not entire, is still very characteristic; consequently if the seat of the cause of the affection were central, we should expect it to cover some ground and cerebral symptoms to coëxist during the period of development.

(4.) An analysis of the case shows the affection to have occurred under circumstances favorable to, and to have developed in the manner of, an abducens paralysis proceeding from external causes.

For these reasons we believe ourselves compelled to regard this recent paralysis of the abducens as of rheumatic origin. Having distinguished between this and the remaining features of the case, let us return to an investigation of the amblyopia.

The ophthalmoscope reveals a senile, ring-shaped atrophy of the choroid around the optic nerve, and besides this a partial, shallow excavation of the papilla, which is not to be regarded as of previous existence (physiological); for in the first place the records, dated at the time of the convalescence of the patient from his apoplectic attack, mention a normal state of the papilla; secondly, the situation of the excavation itself within the papilla is very peculiar. On the right eye, indeed, its place is outwards from the point of exit of the central vessels, extending, however, to the outer edge of the papilla. In the left eye it dips along the inner edge of the papilla, and extends but a short distance beyond the point of exit of the

central vessels, so that the surface of almost the entire outer half
lies in the same plane with the contiguous retina. Its condition in
this eye, therefore, differs materially from that of a physiological
excavation. The state of the case admits of no doubt if we employ
a largely-magnified inverted image and notice the effect of moving
the convex glass.* The edges of the excavation having been once
found, it may also be made out by the increase of whiteness the part
has acquired through the greater prominence of the lamina cribrosa.
Inasmuch as the right half of each papilla is affected by atrophic
excavation, we have to do with a disappearance of nerve-fibres corresponding with the direction of the hemiopia. I lay the more stress
on this result because, with the utmost watchfulness, I never before
succeeded, in a case of cerebral hemiopia, in discovering such a
change capable of being assigned to one half of the optic nerve.
It may be because this occurs so gradually that the proper cases
were not observed for a suitable length of time. As has been already stated, up to the close of the previous record, which was
taken nearly five months after the apoplectic attack, such a discovery
had not been made in the case of our own patient.

The greatest doubt still prevails as to the position in the trunk of the optic nerve
of the fibres pertaining respectively to the fasciculus lateralis and cruciatus. I have
established, I think, beyond all doubt, the old theory of semi-decussation by the
accumulation of exact pathological proof; still it is not yet possible to form a correct
conception of the anatomical situation. If, in a case of perfectly sharply-defined
hemiopia occurring on corresponding sides, the vertical boundary line of the defective portion did not pass through the point of fixation, but rather through the
middle of the "blind spot,"† we should have simply to suppose that the fibres
situated in the outer half of the optic nerve (which radiate outwards on the
retina) belong to the lateral line, those radiating inwards, on the other
hand, to the fasciculus cruciatus. Such a disposition of things, however,
would hardly conform with the functional needs, for then the collective impressions originating in the macula lutea would be transmitted to the cor-

---

* It is hardly necessary to state that this allusion refers to the well-known fact that in
cases of excavation a peculiar ophthalmoscopic effect is produced by using the inverted
image and slightly moving the convex object lens laterally or vertically, keeping always the
same distance from the observed eye. The edges of the excavation are seen to move in a
different plane from its base, seeming to slide over it. This effect is naturally more marked
with an object-glass of comparatively long focal distance, $\frac{1}{3\frac{1}{2}}$ or even $\frac{1}{4}$. It is strange that
no allusion is made to the binocular ophthalmoscope, the use of which throws this method
completely in the shade.—TRANSLATOR.

† Corresponding, of course, in a projected field of vision to the position of the optic-nerve
entrance.—TRANSLATOR.

responding cerebral hemisphere (through the fasciculus lateralis), and we should have to sacrifice the main point in the theory of semi-decussation, according to which the impressions made on identical retinal points are brought together in one tractus opticus, and thus to a centre in one cerebral hemisphere. The fact, that in cases of cerebral hemiopia the line of division passes through the point of fixation, requires a portion of the fibres belonging to the fasciculus cruciatus to radiate outwards from the papilla; in other words, to be already situated in its temporal half. It is not, therefore, in cases of atrophy of the fasciculus cruciatus, to be expected that the atrophy should be strictly confined to the inner half of the optic nerve; while, on the other hand, the paralysis of the fasciculus lateralis cannot involve the entire outer half (reckoning from the point of exit of the vessels). On the whole, however, these collections of fibres make up the bulk of the respective halves of the papilla, and in a case of permanent cerebral hemiopia such a state of the papilla as the present would seem very natural.

The prognosis of the case may be given as favorable, as far as regards the danger of blindness. Even were the apoplectic affection of one side, which has lasted so long, to advance, it would only, in accordance with our theory, cause the hemiopia to become more strongly marked; could not, however, lead to an invasion beyond the vertical line of equal division, or even to a permanent and considerable failure of acuteness of vision. Entire blindness can supervene on a one-sided apoplectic affection only (a) when an apoplectic affection developes itself in the other hemisphere; (b) when fresh effusions into the hemisphere originally affected cause general cerebral derangement, perhaps through anæmia; (c) when a basilar disease, directly affecting the trunks of the optic nerves, supervenes; (d) when a limitation of space in the cranium involves compression of the cavernous sinus, and, in consequence, venous strangulation of the papilla; (e) when the continued progress of the encephalomeningitic disease causes a secondary neuritis. All these possibilities have but little to do with our case. Considering the diseased state of his vascular system, the patient might, to be sure, be seized with an apoplectic effusion in the left hemisphere; such an affection, however, would specially involve the left optic-nerve centre. A fresh and violent effusion on the right side, causing symptoms of general cerebral disease, would either prove fatal or else diminish sufficiently to allow the connection with the left optic-nerve centre to be reëstablished. The purely apoplectic nature of the affection and the absence of all symptoms of implication of the meninges give us at present no reason to suppose any probability of a basilar process. Increase of intra-cranial pressure takes place only in the incipient stage of apoplexy; hence venous strangulation of the papilla, an effect arising only from the prolonged action of such a cause, is not to be seen here, but is especially common with tumors. Derived neuritis (neuritis descendens) finally seems to occur in apoplectic effusions only when reactive prodromal symptoms of a positive character have gained considerably in intensity and extent in the adjacent portion of the cerebrum. In the case of a person who has extensive arterial ossification affecting the action of the heart, and who has already been the subject of an apoplectic effusion, it is of course impossible to foresee the form that later attacks of the same or other affections of the brain may take; this much, however, is certain, viz., that the disease must entirely change its habitation or its shape to produce blindness. A change for the worse (i. e., a more strongly-marked state of the hemiopia) might result from the occurrence of fresh effusions in the right hemisphere, or more extensive changes in the cerebral substance about the previous deposit; still, considering that the condition of things has for three years been entirely *the same*, such apprehensions are of minor consequence. On the other hand, recovery—in other words, restoration of the field of vision—is by no means to be expected. The long duration of this condition of things, and the partial atrophic excavation of the papilla, forbid such a hope.

Be it said in this connection that the resulting on similar sides (i. e., on the left or right side) of limitations of the field of vision or impairments of eccentric vision is very frequent in cases of apoplexy, and that the friends of the patient notice his inability to direct his movements on one side—as, for example, in the taking of things at table. Central vision suffers commonly but little, when the general derangements caused during the period of development are once over, this being equally true of cases of complete cerebral hemiopia, extending to the vertical line of equal division. This latter affection is much more troublesome for scholars when it affects the left tractus opticus rather than the right. For in the first case one loses the eccentric impression of what follows,* so essential to rapid reading, while in the latter it only becomes a little harder to catch a new line after having completed its predecessor.

As regards the treatment, there is nothing to be added. No agent could affect the apoplectic residuum, and we have, therefore, only to

* In the German, "excentrisches Vorauslesen." In reading one word on a page, the main sense of the words immediately succeeding is insensibly perceived, although attention be not voluntarily directed to them. The macula lutea takes cognizance of each successive word, and the lateral portions of the retina simultaneously observe what follows. An exact equivalent of the German phrase it is hard to find.—TRANSLATOR.

counsel an anti-apoplectic manner of life and avoidance of well-known predisposing causes.

In a few weeks the patient was again presented. The paralysis of the right abducens and the consequent diplopia had disappeared under a rather inactive treatment, which fact certainly goes to confirm the theory of a rheumatic paralysis; the derangement of vision was of course the same.

CASE V.

Progressive Amaurosis, coming on under the form of a Central Scotoma, with a simultaneous Anomaly of the Periphery of the Field of Vision.

August N., 23 years old, a robust-looking countryman, presents himself on account of a derangement of vision, which was first noticed in the left eye six months ago, shortly afterwards appeared in the right, and since that time has steadily progressed. We find in the left eye a large central scotoma, with an angle of aperture of about 20ᵃ, within which only bare perception of light exists. Close to the temporal edge of this scotoma the eccentric acuteness of vision is most developed, and diminishes then (the position of the "blind spot" being but a short distance removed from this point) in a nearly normal manner up to the periphery of the field, the sweep of which in the temporal direction is normal. Beyond the nasal edge of the scotoma the field of vision, examined by daylight, seems certainly normal; by subdued lamplight, however, a well-marked indistinctness of the eccentric vision is found to extend to the border of the field, which in this direction is appreciably contracted. Nearly the same state of things obtains upwards, while below the field of vision beyond the scotoma is tolerably normal. The state of the right eye is nearly the same, except only that the scotoma and the anomaly of the field of vision are somewhat less marked inwards and upwards. The patient employs in each eye eccentric fixation, bringing to bear on the object the portion of the retina situated between the fovea centralis and the temporal edge of the optic nerve, as possessing the best lateral vision. He can thus count fingers, with one eye in 6′, with the other in 8′, and with the right can, moreover, decipher the largest letters of the test. During the whole period of development of the disease, and even earlier, the patient has suffered from persistent headache, with a feeling of heaviness, a sense of confusion and occasional giddiness, very marked drowsiness, and in former years from frequent epistaxis; the frontal region, too, on being tapped manifests sensitiveness. In the physical condition otherwise, and the habits of life, nothing of moment is discovered.

The complexion of the case as regards the prognosis is entirely different from that of Case III., where a central scotoma also existed. Our prognosis there was favorable, as far as the danger of blindness was concerned, for the reason that the periphery of the field of vision was entirely normal. But in this case there is, in addition to the scotoma, a considerable contraction of the field of vision upwards and inwards, also an indistinctness of eccentric vision in the same directions. It is this additional fact that causes us to suspect progressive blindness in these cases of central or eccentric interruptions of the field, although with them genuine atrophy seems to be less frequently involved than in the ordinary cases where contractions of the field of vision are alone found (Case II.). In such cases of blindness commencing with scotoma, I have several times had occasion to observe alterations of the cerebral substance following hyperæmia of long duration, and even numerous latent traces of encephalitis. In spite of the bad prog. nosis we associate with the existence of contraction of the field of vision, we will not refer as unreservedly to the necessary approach of blindness as in a case of genuine atrophy (Case II.), for we have in fact here decided symptoms of cerebral congestion, and it is not beyond the bounds of possibility that treatment addressed to them might bring the loss of sight to a stand-still. The patient states that a brother somewhat younger than himself was attacked a few years ago with cerebral symptoms and loss of vision similar to his own, got worse for six months, lost the ability to read, but since that time had remained about the same.

As regards the diagnosis, the persistent congestive headache, especially the local sensitiveness of the cranium, would tend to make us infer the existence of an inflammatory affection, perhaps a chronic meningitis with cerebral hyperæmia, or even an insidious encephalitis. The symptoms, however, do not justify us in pro. nouncing a decided opinion.—The present treatment will be that of chronic meningitis, a "milk-cure," leeches behind the ears, by and bye a seton, and sublimate internally.

This treatment was followed up for several months (Iodide of potash, the decoction of Zittmann and the "cold-water cure" being subsequently employed). The cerebral symptoms disappeared almost entirely, the loss of vision, however, seemed to remain stationary for a time, and then slowly progressed. At the time of the patient's dismissal a small streak of the field of vision still existed, inwards from the central scotoma, leading us to infer that the defective portion at the periphery will speedily be merged in the central portion where vision has failed. In a dim light the patient's movements were already very uncertain. An entirely unfavorable prognosis must therefore be given.

Meanwhile, an examination of the brother of the patient having been made, it was ascertained that while the cerebral symptoms at the commencement of the disease bore the same stamp as in our own case, the physical signs had assumed a different form. The confines of the field of vision had remained entirely normal; the acuteness of vision is now reduced to about $\frac{1}{14}$, owing to an ill-defined central scotoma with an angle of aperture of from 6° to 8°; there is also a moderate amount of atrophic degeneration of the papilla. This state of things has lasted in the present case about four years, and has not yielded to the various remedial measures that have been meanwhile employed. It comes under the head of Case III. Probably both brothers were affected by the same original cause, acting however to a different degree, and exerting consequently a different effect. It has previously been observed that even the benign forms of amblyopia arising from causes connected with the circulation (for example amblyopia potatorum), pass into a form of amaurosis if the cause become more active; and it is especially probable that permanent central scotoma, in the course of which is developed a partial atrophy of the optic nerve, visible on the papilla, needs only the more vigorous action of the same cause to produce progressive atrophy. This undeniable connection of cause should not of course prevent us from making a distinction as regards symptoms in forms of disease the prognosis of which may be widely different.—I have observed hereditary transmission, the possibility of which we cannot deny, less frequently in genuine atrophy than in cases of congestive amblyopia, where the field of vision is either normal or where central interruptions exist; a fact which need not surprise us when we consider the frequent inheritance of a congestive tendency.

Case VI.

Blindness of each Eye of sudden occurrence, with incomplete Restoration of Vision on one side, probably caused by a Basilar Tumor.

Friedrich R., a tailor, 32 years old, pale and bearing evidence of insufficient nourishment, comes to consult us for a recent loss of vision of the right eye and weakness of the left. The functional examination reveals the absence of quantitative perception of light on the right side, diminution of the acuteness of vision to $\frac{1}{2}$ on the left, and (by lamplight) indistinctness of eccentric vision in the outer and lower quadrant of the field. The pupil of the blind right eye is entirely insensible to the influx of light; contracts energetically, however, when the left eye is illuminated, a circumstance which excludes any suspicion of simulation. The optic papilla of each side, as well as the size of the retinal vessels, is entirely normal. The history of the case reveals the fact that the patient during the last few years had had several severe attacks of vertigo, which had twice been so intense as to cause loss of consciousness, and once had induced a temporary weakness of the left arm. The mind remained unaffected, headache occurred only from time to time and very lightly, and the cranium was in no part sensitive. The eyes had been entirely unaffected up to within fourteen days; one morning, however, the patient, while at his work, observed a limitation of the right field of vision, objects far removed to one side seeming entirely to disappear. This obscuration advanced from the temporal side inwards with much regularity, so that on the third day objects at which the right eye was directed seemed situated at the edge of the defective portion. On the sixth day only a faint glimmer of light existed on the nasal side. The day afterwards all perception of light had disappeared. It is only a few days since a decrease in the acuteness of vision of the left eye has been observed, and with this is particularly connected a great sensitiveness to light, indicated by the, so to speak, dazzled manner of the patient.

The method of development of the blindness in the right eye has been emphatically unusual. It differs in its rapid course (in all, six days) from the amaurosis dependent on atrophy of the optic nerve; on the other hand, from anæsthesia of the retina (Case VII.)—to which otherwise the sensitive condition of the patient offers a strong provocation—in the steady advance of the failure in the field of vision, so appreciable to the patient. The measured advance of the contraction of the field of vision naturally gives rise to the supposition of a material agency at work on the trunk of the optic nerve, spreading gradually from the inner to the outer fibres, and inasmuch as a similar effect begins to be produced on the second eye, and no symptoms of a diffused cerebral affection or one existing on both sides are apparent, we must locate this agency at the base of the cranium. The previous attacks that the patient has had are in perfect harmony with the theory of a basilar neoplasma. If the development of such a formation be slow, there may be circumstances under which the adjacent nerves and the cerebrum itself may become adapted to it, allowing of its remaining entirely latent. Compression of the basilar vessels and the derangements of cerebral circulation dependent thereon—for example, faint-like or epileptiform attacks, with, perhaps, interruption of the arterial circulation and transient hemiplegia—occur then only periodically and under the additional influence of accidental causes, and the whole aspect of the disease may for a long time be confined to these symptoms. Paralysis of the cerebral nerves takes place when either the morbid growth, as such, cuts off the nerves, or when the nerve connective-tissue undergoes an irritative process, or when too great an amount of pressure gives rise to compression of the nutrient vessels, and thereby to a loss of nerve-substance. One of these processes must have recently allied itself to the existing difficulty and have affected the optic nerves. The normal condition of all the other nerves, especially of the branches of the oculomotorius, warrants us in placing the morbid growth in front of the chiasma and between the optic-nerve trunks, from which focus its action would first be apparent on the crucial fibres (inner retinal portion). Considering, however, the deficient character of the symptoms, the theory of such a new growth may be set down as the more probable diagnosis. We meet with cases of defined basilar periostitis which, contrary to all expectation, induce no pain, and where the extremely gradual development of the symptoms simulates the progress of a tumor. Finally, there are forms of paralysis which seem to abundantly support the theory of a basilar tumor, but where there is an absolute failure of anatomical confirmation. More prolonged observation will, perhaps, invest such a supposition with increased certainty.

We must give a prognosis unfavorable, to be sure, but not as decided, as regards the vision, as we should in ordinary atrophy. The more the case differs in its aspect from those of frequent occurrence and subjected to the light of abundant experience, the more cautious must we be in our prognostic utterances. If the cause be really a tumor, the final result must be fatal, but a partial restoration of vision is by no means impossible, considering the short duration of the blindness. This would only be the fact in a case where the tumor has really cut off the optic nerve. If the loss of conductive power in the nerves is attributable to inflammatory action in the connective tissue or compression of the nutrient vessels, the processes may be partially transitory and give way to some amount of change. Thus in the case of tumors, in spite of their constant increase, we not infrequently observe an amelioration in certain classes of paralytic symptoms.* For the present it is natural to suppose that the limitation of the left field of vision will continue to progressively develope; the result must show, however, whether it will lead to entire loss of sight in this eye or only to temporal hemiopia.

The patient was again presented, eight days after his admission; the contraction of the left field of vision had meanwhile become more and more extended, following much the same course as did the right, and within a day the patient had been deprived of all perception of light in this second eye, too, becoming, consequently, stone-blind. Moreover, of late, a progressive failure of the sense of smell had taken place, not amounting, however, to complete loss of smell. An ophthalmoscopic examination gives a negative result with regard to the papilla. In the interval a slight attack of faintness had been noted.

The patient remained six days in this state of complete blindness, at the end of which time some perception of light began to manifest itself in the left eye, as well as a gradual restoration of the left field of vision, beginning on the nasal side. Six weeks after the case was first presented, the field of vision of this eye had attained a nearly normal development; the acuteness of vision, however, had only reached $\frac{1}{15}$, and eccentric vision in the neighborhood of the temporal edge remained indistinct. This improvement and simultaneous increase of the sense of smell followed the administration of the lactate of zinc in increasing doses; was probably, however,

* See the instructive case published by Sämisch in the February number of this year's magazine, pp. 51–55.

not connected with this agent, to which, in cases of idiopathic retinal hyperæsthesia (see Case VII.) I am highly partial. The right eye remains entirely blind, and its papilla now offers clear evidence of atrophic degeneration.

The last recorded condition of things seems now (a month later) to remain unaltered. That the partial restoration of vision on the left side by no means detracts from our original diagnostic supposition of a basilar neoplasma, has already been settled. The patient, too, continues pale and exceedingly decrepid; the left papilla, in its turn, commences gradually to exhibit traces of atrophic degeneration, without, however, any reduction in the partially-recovered vision.

CASE VII.

Anæsthesia of the Retina, with Concentric Limitation of the Field of Vision; Quick Recovery.

Carl S., a delicate boy, 10 years of age, is brought to the clinique for deranged vision of the right eye and twitching of the face. The right eye is extremely intolerant of light, it being hardly in his power to hold it open when exposed to strong daylight; on softening the light, however, and neutralizing a hyperopia $\frac{1}{30}$, the acuteness of vision is found to be $\frac{1}{3}$; the field of vision is concentrically, though irregularly limited, rather more downwards than in any other direction. Its angle of aperture in the vertical direction is about 40°; in the horizontal, 50°. By lamplight, even though limited in amount, no diminution in this diameter of the field of vision is noticed; in fact, it rather increases, and this, too, is the case when the patient is made to look by daylight through dark-blue glasses (shade No. 8). *In every direction phosphenes are producible.* It is particularly striking that pressure behind the upper part of the ora serrata at once brings out the lower phosphene, although the transmission of impressions is most deranged in this direction. The phosphene in question is projected at least thirty degrees below the edge of the contracted field of vision. The result of the ophthalmoscopic examination is entirely negative. The acuteness and field of vision of the left eye are normal. On the right half of the face periodic twitchings of separate muscles occur, particularly of the zygomatici and levatores; which increase, it is true, when the admission of bright light causes the right eye to be closed, but last even when it is entirely shaded. The interval between their occurrence is seldom more than half a minute, the twitchings themselves being slight and lasting only a few seconds.

With the exception of some nervous irritability, the patient's health was always good. Three weeks before he presented himself he was, while walking in the country, caught in a thunder storm, and much frightened at a tree, a short distance from him, being struck by lightning. The next morning, the derangement of vision and the twitchings were both observed.

The presentation of these particulars was coupled with the following remarks :—We have here such a case of partial anæsthesia of the retina, especially of its peripheric zone, as often occurs in excitable children and nervous or hysterical women, and inspires even experienced ophthalmic surgeons with an erroneous fear of progressive amaurosis. The particular cause seems peculiar in such cases; it is, however, probable that we have here simply to regard the accompanying mental impression. Such forms of anæsthesia, coupled sometimes with a loss of cutaneous sensibility to pain, or, as in the present instance, with twitchings, may be seen to particularly occur in those cases where general excitability has acted as the predisposing mental agitation, as the immediate cause. They therefore particularly affect individuals of excitable temperament, the subjects of anæmia, those convalescing from severe diseases— for example, children getting over scarlet fever, measles and typhoid, whose power of resistance has not yet become developed. In comparing the characteristic signs in these cases with those in amaurotic affections, we shall find the former as follows :—

(1.) Only a slight diminution of the central acuteness of vision, seldom passing the bounds of $\frac{1}{4}$ or $\frac{1}{4}$, while there exists an important anomaly in the confines of the field of vision, generally consisting in concentric, irregular limitation.

(2.) A development either sudden or reaching its acme in a few hours or days.

(3.) Retention of the phosphenes at points corresponding to these portions of the retina where there is no perception of light, showing a loss of connection between the rods and fibres, dependent on a local cause.

(4.) Simultaneous hyperæsthesia of the retina and the active character of the retinal difficulty thereon dependent, in consequence

8

of which vision either remains as good or improves on using dark-blue glasses or in a dim light, circumstances which in general cause a diminution of vision.

(5.) The age and sex of the patient. It is well known that the amaurotic affections dependent on atrophy of the optic nerve—leaving out of the question congenital states as well as further intracranial derangements, manifesting their existence by palpable symptoms—very seldom occur with children and, in the case of adults, are infinitely more frequent with men than women. In this form of retinal anæsthesia the opposite is the case. According to my observations it affects almost exclusively women and children, and in the exceptional cases where men were seized — a thing that has happened only twice in my experience — the subjects were those where temperament and bodily constitution approached either the feminine type or that of the child.

Finally the predisposing and accompanying circumstances are of importance, because under the first head we mostly meet with mental impressions, under the second with a loss of cutaneous sensibility and local affections of the motory system. In our own case a preliminary examination of the sensitiveness of the skin has given only a negative result. On the other hand the peculiar facial affection is completely characteristic.

As a whole these points of differential diagnosis are not without value, though taken separately the departures from them may be numerous. And first we have those exceptional cases in which (contrary to No. 1) the acuteness as well as field of vision suffers an unusual diminution. I can on this occasion refer to two in which blindness had nearly been produced. An unfavorable prognosis might have been given, still the other symptoms seemed to justify me in speaking encouragingly, and the usual treatment was followed by entire recovery. It is possible that certain cases of sudden and entire blindness might also be included in this category; it is not, however, in our power to establish rules of differential diagnosis between them and other incurable forms.

In the case of a boy 8 years of age a very remarkable phenomenon was observed. After a concentric limitation of the field of vision had lasted some time, and all other symptoms had decidedly indicated that the case belonged to the foregoing category, the hyperæsthesia of the retina very much increased, and the next day there took place entire restoration of the periphery of the field of vision, accompanied, however, in each eye by a large central scotoma, diminishing the acuteness of vision from $\frac{1}{2}$ to $\frac{1}{40}$ (eccentric). This turn so surprised me that I at first refused credence to the statement, until over-persuaded by measuring the scotoma at various distances and by following up the case. The boy was convalescent from the measles, always delicate and possessed much nervous excitability; the usual treatment brought about his cure.

The amount, too, of the accompanying retinal hyperæsthesia varies extremely, being often extremely marked in the case of hysterical patients, while children between 6 and 14 years of age, the most frequent subjects of the disease, may have only a moderate feeling of being dazzled. The affection of one side only makes the above-cited case exceptional; the disease almost always occurs on both sides, though it may be to a different extent. The fact that in cases of recent occurrence the optic papilla retains its normal appearance is of course of no value as regards the differential diagnosis, inasmuch as this is also applicable to cases of a serious nature that have existed but a short time (Case VI.). On the other hand it is remarkable that even where the difficulty has existed some time and not been treated, the papilla may retain its normal redness, transparency and superficies.

It follows from what has been said that we hold the prognosis in these cases of retinal anæsthesia to be favorable. An entire cure is generally effected within a few weeks; occasionally the disease remains for some time at a certain height till the advent of convalescence, and only in a few cases have I observed the cure notwithstanding a long delay to be incomplete, owing to the but partial disappearance of the peripheric contraction of the field of vision and of the hyperæsthesia — of the last, especially, where (as in cases of hystoria) the general health could not be established on a firm foundation. I have never observed these cases to result in amaurotic blindness.

As regards the question of treatment much stress is first of all to be laid on the regulation of the allowance of light. The good results that have been said to follow the entire and methodic deprivation of light in amaurotic affections may, I think, be attributed to the fact that the cases were either those in point or else of the nature of hemeralopia, a state in many respects analogous, different though it be with regard to the retinal torpor. In progressive atrophy the light, it is true, should be softened in order to remove a cause that may accelerate blindness; this is, however, never observed to produce a remarkably curative effect. It is my custom to first keep those affected with anæsthesia of the retina for several days in a completely darkened chamber, and then, during perhaps 6 or 8 days, to allow a gradual increase in the amount of light. Later, when the patients are allowed to go out, blue glasses of different shades, varying according to the degree of light they are exposed to, are to be given them. The importance of this portion of the treatment varies directly with the amount of retinal hyperæsthesia.

Among medicinal agents I place my chief reliance on the internal administration of the preparations of zinc in increasing doses, following the method recommended by Jaksch in the treatment of a loss of cutaneous sensibility. I formerly frequently employed tartar emetic in nauseating doses, and generally obtained a satisfactory result; this plan, however, is far more disagreeable to the patient than the treatment by zinc, and where the digestion is affected cannot be used indiscriminately; I therefore counsel that it be had recourse to only when the use of zinc has failed to produce an impression. If improvement has once commenced, I go over to mild tonics, administer iron, give aromatic and salt baths and cold spongings. All this treatment must have reference to the state of the general health in the case before us. It is indubitable that a well-balanced mind exercises a decided influence. As the passions often furnish the exciting cause, so have I seen their indulgence followed by a relapse in many cases where a cure was already in progress. Setting the patient at ease as to the true nature of the affection, often brings about the favorable crisis. During treatment the exercise of the accommodation must be entirely forbidden; on the other hand when the dark chamber can once be dispensed with, much time should be spent in the open air. In this form of disease I must caution against the abstraction of blood, as well as all remedies which reduce the system, excite the nerves or interrupt sleep. Their employment is not only followed by an immediate change for the worse, but the disease developes a more obstinate character.

The boy was subjected to such a course of treatment, the light being regulated and zinc administered (lactate of zinc, at first gr. iss., afterwards gr. v. a day). Twelve days after his admission the twitchings of the right half of the face had been reduced to a minimum, the acuteness of vision had increased to more than $\frac{2}{3}$. The limitation of the field of vision outwards, inwards, and upwards had disappeared, and was only perceptible downwards; the retinal

hyperæsthesia had diminished, but was not entirely obviated. The
patient was now ordered iron and cold spongings, and, when next
shown at the clinique, the ability to use the eyes freely and a *resti-
tutio ad integrum* were demonstrated, the cure having taken in all
four weeks.

CASE VIII.

*Temporal Hemiopia, following a Basilar Affection (supposed Perios-
titis); Dubious Prognosis; Recovery.*

Mrs. Emily B., 36 years of age, presents herself at the clinique
on account of an impairment of vision that has lasted only eight
days. On examination, there is found in each eye acuteness of
vision of only ½, as also an entirely symmetrical defect in each temporal half of the field of vision. All perception of light is
lost over the space extending outwards from a line going nearly
through the middle of the "blind spot," while over the space between such a line and a vertical one passing through the point of
fixation, perception is so imperfect that fingers can be counted only
in the immediate vicinity of the latter. On the nasal half of the
field, on the contrary, eccentric vision is everywhere normal, even by
reduced lamplight. In conformity to this the temporal phosphene
is entirely wanting, while the nasal is produced with great ease and
distinctness. The ophthalmoscopic examination gives an entirely
negative result. As regards the antecedents and the mode of development, the following is obtained: the previous health of the patient had always been good, and no trace of syphilis had ever existed; but several months after her last (seventh) delivery she had
been seized with headache of extraordinary severity, and at the
same time with diplopia. She had been at the clinique for this half
a year before her present visit, and at that time the cause of the
diplopia was found to be a paralysis of the right abducens, the
functional state of the retina itself seeming to be entirely normal.
The cause of the paralysis, as well as of the headache, we supposed
to be akin to periostitis (basilar). The headache entirely yielded
at that time to a derivative course of treatment, but the paralysis of
the abducens proved excessively obstinate, so that after iodide of
potash had been given for several months and electricity tried, a loss
of mobility to the extent of more than 1½‴ remained almost stationary, and the consequent confusion of vision (contraction of the

internus having resulted) had to be relieved by setting back the insertion of the left rectus internus. From that time up to within eight days of the present visit the patient had been perfectly well; the catamenia had, however, failed to appear at the last two periods, and she was then attacked by heavy and violent pains extending over the entire head, as well as by the derangement of vision already referred to, which continued steadily to increase.

No particular result is obtained from a general physical examination. The patient is free from fever, but has a worn look, attributable, no doubt, to the loss of sleep caused by the pains in the head. Neither is anything definite discovered by the exploration of the orbits; each bulbus allows itself to be pressed against the cushion of fat without thereby exhibiting any signs of sensitiveness. On the other hand the skull, particularly on the level of the basis cranii, is exceedingly sensitive to tapping, the most pain being caused when two points, situated opposite each other, are simultaneously tapped.

It is clear that the reason of the present derangement of vision must be located at the basis cranii. No change that could account for the limitation of the field of vision is to be met with in the eye; such cases of temporal hemiopia being in general hardly ever dependent on intra-ocular complication. An orbital cause might certainly exert an influence on the nasal portion of the optic nerve; still, leaving out·of account the fact that such a cause must be symmetrical and affect both sides, there exists no sort of basis for such a supposition. On the other hand, the localization of the difficulty at the middle of the basis cranii satisfactorily explains all the symptoms. We know that a cause of disease which has its seat here and acts upon the optic nerves, first affects the fasciculi cruciati, and thus involves the connection with the temporal border of the field of vision. If the action extend equally on the two sides, the temporal limitation of the field of vision will be symmetrical. We never in these cases find the defective portion standing out in such sharp relief against the part that retains its normal functions as in hemiopia on similar sides. This is, à priori, comprehensible. When one tractus opticus is paralyzed, perhaps from a loss of cerebral connection, the boundary of the perceptive portion becomes sharply defined, corresponding to the distribution of fibres of this tractus on the retina, and following the example of what occurs in paralysis of the trigeminus, in the median line of the face. If, on the contrary, the two

optic tracts feel the influence of a morbid cause having its seat at
the middle of the basis cranii—an irritated state, for example, of
the connective tissue, proceeding from the periosteum—it would be
hard to understand how the action thereon dependent should expend
its full intensity on certain bundles of fibres without to some extent
implicating those adjoining. We therefore find under all circumstances in cases of temporal hemiopia, an intermediate region* in
the field of vision. The case of the patient now introduced completely harmonizes with this view—the field of vision, as already
stated, failing entirely beyond the "blind spot," while between the
"blind spot" and point of fixation a gradual increase of sensitiveness to impressions takes place. The theory of a basilar origin
finds further support in the nature of the pains in the head and the
increase in them caused by tapping the head in the region of the
basis cranii, a symptom which, though wanting in many cases of
basilar disease, is certainly not without significance when present.
A similar meaning is to be ascribed to the nature of the affection
passed through six months ago. Besides intense and general headache,
it was characterized by the particular obstinacy of the accompanying
paralysis of the abducens. Experience teaches us that the ordinary,
so-called rheumatic paralysis of the abducens may be accompanied
at the time of its development by local pains in the forehead and
temples, but not by general and severe headache; moreover, when
once the greater part of the power of motion has been restored,
such cases of paralysis generally recover more uniformly and completely than was the case here. Finally, if the location of the difficulty within the cavity of the cranium be once allowed, the theory
of a basilar origin receives support from the fact that there is not
the slightest reason for supposing it to be in the cerebrum. In spite
of the pronounced paralysis, formerly of the abducens and now of
the crucial fibres of the optic nerves, there have been neither hemi-
plegic attacks, mental derangement, nor affections of the head of
any kind to indicate a disease of the cerebral substance.

* *Uebergangsbezirk.* No analogous terms exist in the English language for many similar
expressions. We are here to understand that the portions of the retina respectively des-
titute of vision, and which retain their normal powers, are not sharply separated from each
other by a defined boundary, but slowly merge, the one into the other, over a region
which, if the term be literally translated, bears the name of the "district of transition."—
TRANSLATOR.

The task of deciding the nature of the basilar cause is much more difficult. The rapid development of the symptoms supervening on a state of perfect health, the extreme severity of the commencing headache, the entire intermission between the two attacks, seem at first sight to indicate an inflammatory condition rather than a new growth. Still, the possibility of the latter must not be entirely excluded. When tumors gradually form at the basis cranii, they may, as such, remain latent, and only betray their presence periodically by taking on a state of irritation. On the other hand, it is the exception for new growths that have once given rise to symptoms of paralysis, to allow a temporary return of an entirely normal condition. As a rule, we have only variations in the symptoms of paralysis, some taking their departure and others making their appearance or persisting. When we are not able to form a definite diagnosis with certainty, and probabilities are evenly balanced, it is without doubt a sound, practical principle to proceed on that theory which seems to open to us the best field for effort. This is preëminently true when the probabilities incline in favor of that theory. Let us then, in the case of our patient, for the present dismiss the idea of a basilar neoplasma, and give our attention to the theory of an inflammatory affection. And this can be well located only in the dura mater. Inflammations of the more delicate cerebral membranes have too great a tendency to diffusion to cause paralytic affections within such narrow limits. Their development is generally rapid, and is accompanied by febrile and general cerebral symptoms, while they pass off in a different manner. The date of the first attack renders it, moreover, possible for us to connect the defined pachymeningitis, which we feel warranted in assuming, with the puerperal state. We possess relatively few anatomical and clinical facts with regard to such defined basilar affections. I have, however, acquired the conviction, in which, too, the results of autopsies·have strengthened me, that the most varied forms of basilar paralysis spring from this source. This is particularly applicable to certain forms of recurring paralysis of the muscles of the eye, in fact processes of a periostitic nature are as a rule prone to recur.

Our prognosis cannot be other than doubtful. The difference in this respect between temporal hemiopia and that occurring on similar sides (Case IV.) is very striking. While in the latter the continued action of the same morbid cause does but complete the hemiopia,

and never causes blindness of either one or both eyes, it is of course
possible for a source of disease situated at the base of the brain to
exercise a constantly increasing effect on both optic nerves, overstep
the limits of the fasciculi cruciati, and end in absolute obliteration
of the field of vision (Case VI.). On the other hand, an entire
pause in, or even complete disappearance of, the disease may occur
at any stage whatsoever. This would substantially depend on the
nature of the morbid cause. Inasmuch as in our case the supposed
cause may disappear, and has been so short a time in force that dis-
integration of the nervous elements is neither to be expected, nor
visible on the papilla, recovery is possible. A more exact definition
of our prognosis, as regards either the derangement of vision or the
subsequent result, must naturally be based on the course of the
disease.

The patient was first subjected to a derivative plan of treatment
(leeches behind the ears, then dry cupping, drastic pills, derivative
foot-baths, later iodide of potash). Under this the headache entirely
ceased, but the field of vision remained the same, and its acuteness
even became reduced to ¼. The general condition, too, excited constantly increasing fears. The urine became considerably increased
in quantity, and marked pallor, loss of flesh and weakness simultaneously presented themselves. The daily amount of urine was from
4000 to 6000 cubic centimetres; its specific gravity varied between
1002 and 1005. The color of the urine was extremely light; it
was examined by Dr. Kühne, and proved to contain neither sugar
nor inosit. In the morning, thirst became unquenchable. These
symptoms having reached their height about four weeks after her
admission, and the bodily weight having fallen, without any increase
of temperature, to 94 pounds, the dose of iodide of potash was reduced from ℈ i. to ℈ ss., and liquor ferri chlorati at the same time
ordered. A few days after this prescription, but possibly entirely
independent of it, a diminution of the thirst and in the quantity of
urine showed itself, shortly after which the bodily weight and vision
commenced steadily to increase. Seven weeks after her first introduction no defect in the field of vision could be made out; the
eccentric vision, however, in the temporal zone, particularly outwards
and downwards, was still indistinct; the acuteness of vision amount-
ed to ⅔; weight 100 pounds; average quantity of urine in the twenty-four hours rather over 2000 cubic centimetres; specific gravity,

9

1010. Four weeks later the patient was discharged entirely convalescent; weight, 108 pounds; urine normal in amount; field of
vision irreproachable, even by diminished light; acuteness of vision
more than ⅖; satisfactory complexion and tone.

I would state, in conclusion, that the patient, who is a resident of
this city, has been shown at considerable intervals, and that the last
extract from our records, written more than a year after the derangement of vision, bears witness to an entirely normal condition
of things. The catamenia, too, reappeared during the progress of
the convalescence. The entire and apparently permanent recovery
gives, we think, an increased support to our original diagnosis of a
local basilar periostitis, although, of course, the obscurity of these
regions leaves ample room for doubt. I was particularly interested
in this case, partly on account of the entire disappearance of so
marked a hemiopia, partly because the urine, in spite of its increase
in quantity, contained neither sugar nor inosit. As regards the first,
my experience permits me to recal but few such fortunate results;
the coincidence with intra-cranial processes of an increase in the
quantity of urine has been, it is true, considerably studied; but, as
far as I am aware, not yet recorded in connection with such a group
of symptoms as the present.

CLINICAL REMARKS ON A CASE

OF

EXTRACTION OF CATARACT.*

Delivered Jan. 3d, 5th, 6th, 7th, 9th and 20th, 1863.

MRS. D., 64 years old, has a hard, ripe cataract in the right eye, an unripe one in the left. The external appearance of the eyes indicates no complication. The cataract of the left being unripe, fingers can be counted in 2', while the perceptive power of the right eye has been measured by means of our "graduated diaphragm." If we use an entirely darkened chamber, placing the diaphragm 8" (the distance which for the sake of uniformity we always employ) from the eye to be examined, and reduce the intensity from 100 to 1, the patient gives entirely correct answers till 4 is reached, but at 2 begins to be uncertain. This is exactly what we should expect when the cataract is ripe and saturated, as is here the case. If the diaphragm be set at 25 and moved from side to side, the patient follows it with the eye as accurately as could be expected, considering the diffusion of light. The question as to whether any considerable near-sightedness had previously existed, a question which—in view of the frequent

---

* From a letter of von Graefe to the Editors :—

"The case which forms the subject of this lecture is by no means an exceptional one. I am, however, of opinion that a description of those irregular complications that may occur during the healing process after flap-extraction, as well as an allusion to various points that are now, in this connection, receiving special attention, may be of use to the practitioner; and do not hesitate, therefore, to offer some such material to the 'Klinische Monatsblätter.' The most assiduous observation and the most comprehensive experience avail of course only up to a certain point in answering the question why flap-extraction is sometimes followed by unfavorable results, just as the problem in general surgery, as to why wounds, that show every disposition to unite by first intention, sometimes suppurate, is capable of an only partial solution. Still we may attempt to approximate to it by carefully noting and comparing the different circumstances influencing the patients, the eyes, the operation and the after-treatment in cases resulting favorably and unfavorably. And although I reserve for a special volume the results of my own experience in this connection, embracing at present 1500 cases, I am still disposed to take a few cases of flap-extraction to illustrate some clinical remarks bearing on this subject."

combination of cataract with *sclerectasia posterior*—should never be omitted, was also answered in the negative. In a word, we have before us a so-called *cataracta simplex.*

I say intentionally *so called ;* for strictly speaking it may be said that all eyes, in which cataract is spontaneously developed, may be found to have undergone more or less change in other respects, particularly in the hyaloid membranes and vascular system : in fact, these changes may well come to be looked upon as the pathological point of departure of the cataract ; meanwhile, however, the very great differences in respect to recovery, which at present fill us with astonishment and make us almost lose confidence in ourselves—especially when we compare them with the results of traumatic affections of otherwise healthy eyes—these differences, I say, may be explained by a closer study of the nature of the anomalies which are at the bottom of the formation of the cataract. Such theorizing, however, carries us into an obscure future ; it suits our present purpose to limit *cataracta simplex* to those cases where we are unable to discover any affection of the deeper structures of the eye, and no break exists in the nervous apparatus.

Shall this patient be operated on at present ? There is certainly no valid reason for delay ; in fact, we hold it prudent to do the operation of flap-extraction on only *one* eye at a time,* and it is only in exceptional cases and under extraordinary circumstances that we depart from this rule. In one eye the cataract is ripe. Did the patient enjoy tolerable vision with the other eye, and could she gain her living, we might advise a postponement of the operation; for three months, however, she has not been able to go about alone, which fact makes the recovery of sight a matter of prime necessity under any circumstances.

Let us now come to an understanding with regard to the *prognosis of a flap-extraction* in the present case. And first, as to the general state of the system. The patient, it is true, is only 64 years old, but prematurely marasmic. This is sufficiently shown in the deep wrinkles, the atrophy of the skin all over the body, particularly on the neck and backs of the hands, the scanty and thinly sown gray

---

[* In view of the difference of opinion on this important point, which has prevailed and to some extent still exists in this country, the translator would call attention to the accompanying passage from the excellent work of Wecker, now in process of publication :—

"The elementary principles of prudence indicate sufficiently the impropriety of performing the operation for cataract on more than one eye at a time. In the first place, the consequences of the first operation and the results which it furnishes are, for the surgeon, the source of valuable information with regard to the line of conduct he should pursue in the second. In the second place, no conscientious operator would consent to expose his patient to the risk of losing, at one cast, all hope of recovering sight, a thing which might happen after a double extraction, followed by accidents entirely independent of the operation itself, and determined by the imprudence of the patient or of the persons surrounding him."— *Etudes Ophthalmologiques,* tome ii. p. 255.]

hair, the somewhat stooping form and a weight of 88 pounds to a height of 5 feet 3 inches. In my estimation, marasmus should exert a most unfavorable influence on the prognosis of flap-extraction, that which is premature more than that which legitimately belongs to the time of life.* But we find, moreover, in the eyes themselves circumstances which contribute to depress our hopes, namely, the deeply sunken position of the balls, the small corneal diameter, which we have just found to hardly measure 4¾''', and a tremulousness in the muscles when the eyes are turned far in or out. A sunken position of the balls has been particularly dwelt upon as an unfavorable circumstance, in so far as it interferes with the mechanical execution of the operation. Viewed in this aspect, I do not attach much importance to it, for when extraction is done downwards a moderate amount of practice enables us to overcome the difficulties in question. But an abnormally deep position of the eyes, depending, as it does in general marasmus, on the disappearance incidental to age of the fatty tissue of the orbit, is, it is somewhat probable, an unfavorable sign as regards recovery; in fact, it is very probably so when, coupled with it, is the second sign to which we have referred, the senile diminution in diameter of the cornea. Where such a combination exists we are almost sure to have collapse of the cornea after the operation; that peculiar form of collapse in which the cornea not only falls in, lies loose and in folds, but also exhibits a concentric shrivelling, the whole indicating that the corneal tissue is capable of but little resistance. Experience shows more and more that the loss of nutrition which is brought about by the flap-section is fraught with peculiar danger to such a cornea, and is followed not infrequently by total or partial suppuration. The same thing happens in the case of these thin and but slightly elastic corneæ, as in a plastic operation where the skin is thin, atrophic, and devoid of panniculus; a loss of nutrition that, under more favorable circumstances, would be perfectly supportable, giving rise to fears of suppurative destruction or entire necrosis, an unusual amount of contraction being observable in the skin as soon as separated from its original insertion. Finally, the third sign, viz., the nystagmic trembling of the muscles

---

* The opinion of some colleagues that extreme age does not exert a particularly unfavorable influence on extraction is most emphatically contradicted by my tables, which show after the 65th, and particularly after the 70th year, a considerable falling off in the percentage of recoveries.

when made to exert their utmost contractile power, goes to show a senile atrophy of the muscular apparatus and to confirm our fears that the wound may not heal well. *It is thus made evident that the danger of a diffuse (necrotic) as well as of a defined (flap) suppurative process is extraordinarily great in the present case.*

The tendency, too, to *iritis* is decidedly greater in marasmic eyes (increasing, as it does, in proportion to the age). The hardness of the cataract tends to render this more probable, greater violence being done as it passes through the pupil. Still, I should not say that the affair promised particularly ill in this respect; for one of the cataracts is entirely ripe, from which we infer an entire and easy separation of it from the capsule, and have neither to apprehend difficulty or delay in the removal of the cortical masses, nor that any will remain behind—a fertile cause of iritis. Moreover, atropine brings about a complete dilatation of the pupil, a small border of iris alone remaining, and it continues well dilated for four or five days, a thing by no means universal with old people, and which indicates relatively less tendency to iritis. In order not to be misunderstood, I should here add that I consider the iritis following extraction as capable of classification under two heads—*transplanted* and *genuine.* The transplanted follows upon or becomes developed with the suppuration of the wound, the latter either leading to a cell-growth, on the posterior surface of the cornea, which pushes into the pupil, or else the suppuration seizes upon the portion of the iris in the immediate vicinity of the wound, and thence invades the remainder of the iris and the ciliary region. The genuine iritis, which seldom comes on before the third day, and generally makes its appearance between the fourth and tenth, has no direct effect on the healing of the wound, but when of early occurrence may influence it unfavorably, though indirectly. This form alone was, of course, that to which we had reference, the transplanted variety having its own exciting cause.

Finally, it should be stated that the patient suffers from an old stoppage of the lachrymal passage and moderate epiphora, a circumstance which is shown from experience to somewhat diminish the chances in favor of the healing of the wound.

Taking all these circumstances into account, the general prognosis of flap-extraction must be here essentially modified. According to my reckoning, of a hundred cases of flap extraction 65 result favorably, by which I mean the gaining an acuteness of vision of at least

$\frac{1}{4}$; if more than 75 years of age, at least $\frac{1}{6}$. In 15 of the remaining 35 a favorable result is attained by a subsequent operation, consisting either in an operation for secondary cataract, or in an iridectomy with an operation for secondary cataract; of the 20 that now remain about a third get at least vision enough to go about alone (acuteness of vision $\frac{1}{50}$ to $\frac{1}{30}$), a second third gain still less, and from 6 to 8 per cent. of all eyes operated on remain or become entirely blind; that is, deprived of all ability to distinguish objects (whether they have quantitative perception of light or no). This is the final exhibit, when I take into account all the cases of *cataracta simplex** where an operation seems indicated.

Under other circumstances much more flattering statistics could be shown; my own, for instance, would have been twice as good if I had only included the operations on the occupants of the private rooms of my infirmary, and omitted the poor patients, the majority of whom were operated on under very unfavorable circumstances. Far more favorable results, too, might be furnished by those practitioners who either employ another kind of operation in unfavorable cases, or decline such cases altogether, than by those who perform the operation of flap-extraction in all those cases in which an operation is not contra-indicated by the tenets of our science. Finally, I am convinced by my own experience that the recovery after extraction, like that after all surgical operations, is influenced by differences of climate. I therefore lay only a relative stress on the above statistics, and communicate them solely in order to furnish an average scale to the less experienced practitioner, who may wish to make a truthful statement to his patient of the chances of the undertaking; for the sake, moreover, of assigning to the present case its individual place on the general scale of prognosis.

As may be readily seen, the prognosis of extraction may be infinitely better in a single case than in the .long run, and *vice versa*. Imagine a patient in the vicinity of fifty years of age, perfectly healthy, of an equable temperament, submissive, hopeful as regards his future, with a large cornea of $5\frac{1}{4}'''$, a cataract ripe for the last quarter or half year, and with soft cortical substance—the chance of success would here be exceedingly good, infinitely better than the given average. An opposite condition of things inclines the scale in the contrary direction, and thus is it in fact in the case before us. I should here hardly estimate the chances of immediate and entire

---

* The more marked cases of myopia were formerly excluded from the list; latterly, however, have been reckoned in, provided no amblyopic complication could be discovered before the operation. This seemed allowable because, contrary to anticipations, experience has shown the healing process to be in no wise unfavorably affected by the staphyloma posticum (provided the vitreous is not considerably changed).

success as more than even, while the chances of an ultimate recovery, to be obtained by a secondary operation, are as 2 to 1.

Is, then, for this reason, another plan of operating to be selected? The hardness of the cataract forbids the thought of discission, an operation, indeed, seldom advisable in advanced age; nor are we likely to choose linear extraction, with or without iridectomy, an operation which, in cataracts as consistent as the present, does too much violence, leaves cortical substance behind, and is followed by chronic iritis, which generally invades the ciliary region. Our sole choice, therefore, lies between flap-extraction and reclination. The chance, on the whole, after reclination is very unfavorable compared with that after flap-extraction. If the average success following the former is to be reckoned as at least 80 per cent., that belonging to the latter is at the outside 50 per cent. Still, a fair field for reclination might be found in those cases where the individual circumstances cause the chances of extraction to fall considerably below the average. It is, however, in this connection to be regretted that most of the physical objections which, in a given case, apply to flap-extraction, may be urged with equal or greater cogency against reclination. Thus it is, for example, with certain complications which have been actually brought forward as contra-indications to flap-extraction, for example chronic choroiditis with a fluid condition of the vitreous. Their existence, of course, interposes a serious obstacle in the way of extraction, but a still greater one in that of reclination. Supposing, then, an operation to be advisable, they are not to be considered as contra-indications to the former, but simply as unfavorably modifying the prognosis. The same is true, though perhaps not so emphatically, of marasmic eyes like the present, unable as they are to long endure the irritation set up by the displaced lens, the tendency being to the development of chronic cyclitis or deep-seated inflammations, attended by increased secretion and ending in excavation of the optic nerve, the whole contributing to establish a prognosis for reclination much below the average. In my opinion, reclination is to be regarded as an incontestably proper though exceptional procedure in those cases where the general state of the system renders the healing of the wound doubtful, and where the danger of choroiditis or of inflammation with hypersecretion does not increase in the same ratio. After all, a too exclusive predilection should not be allowed to turn the scale in balancing chances. Signal as seems in the light of today to be the blunder of submitting to the insidious operation of reclination an eye which offers an average or even better prospect of success from extraction, there yet remains scope for the exercise of some choice or preference on the part of the operator in cases like the present, where extraction offers a decidedly diminished chance. And although in the case of our patient extraction seems on the whole more advisable than reclination, we should readily comprehend and in no wise find fault with the course of any colleague who employed by preference the needle.

Flap-extraction having once been determined on, we have to decide whether to *precede it by an iridectomy*, or perhaps combine the two operations. What advantage is to be gained by combining an iridectomy with a flap-extraction? Does it ward off the danger of *diffuse suppuration of the cornea*, a process occurring as a general thing between twelve and twenty-four hours after the operation, and characterized by profuse secretion, appearances of swelling and the rapid formation of a purulent infiltration encircling the entire cornea, and premonitory of necrosis of this structure? Not in the least. We have seen this very thing occurring in the same manner and running the same course in cases of extraction where the precaution of making an artificial pupil had been adopted. Only in proportion as the presence of the coloboma renders the mechanical execution of the operation in itself more easy, can it be said that an indirect influence has been exerted on the occurrence of the above process.

Does the presence of the coloboma afford any protection against *defined suppuration*, the symptoms of which generally make their appearance somewhat later (eighteen to thirty-six hours), are otherwise externally similar to those of diffuse suppuration, except that the secretion is less abundant, and after having been poured out for the first time somewhat diminishes, in which, however, the suppurative process confines itself to the vicinity of the wound or the corneal flap, and at the most exhibits a tendency to send a ring-shaped infiltration into the uncut portion of the cornea? This query, too, must be answered in the negative. Since the time that I have combined iridectomy with extraction to meet various indications, I have not noticed any influence of the procedure (making again the same allowance as above for its effect on the act of the operation) on the *occurrence* of defined suppuration of the wound, but have, however—and this is a very important point—on its *course*. The principal

70

danger of defined suppuration, always supposing it does not ultimately become diffuse, lies not so much in the destruction of the cornea, as in the iritis transplanted from the wound (see above). The masses of pus make their way into the anterior chamber, the iris becomes the seat of suppurative swelling, and this transplanted suppurative iritis exhibits a peculiar tendency to invade the ciliary region, and thus lead to irrecoverable loss of sight from ciliary exudations and atrophy of the bulb. The existence of a coloboma of the iris does not, it is true, remove the possibility of such an invasion, but is undeniably efficacious in hindering its progress. The process attains a diminished height, union of the edge of the pupil with the capsular cavity takes place more quietly and frequently to a less extent, and purulent cyclitis fails to make its appearance in many cases, where, but for the coloboma, it might have been expected.

Does the iridectomy afford any protection against *general iritis*, the development of which is in most cases owing to contusion of the iris during the operation, or to the leaving behind of cortical remains? To a certain extent, we need not hesitate to answer this in the affirmative. The protection furnished is, of course, by no means absolute; genuine iritis, however, is less frequently observed where iridectomy has been done in advance, and when it does occur is, as a rule, less severe. It is plain that the amount of contusion inflicted during the operation is less when a coloboma of the iris exists opposite the apex of the flap. And, where the amount of contusion is the same, an iris in which a coloboma has been made exhibits a lessened tendency to inflammation, and inflammation of it is less dangerous; on all sides, therefore, are favorable circumstances which give us an *a priori* ground for foreseeing the result which experience renders secure. It is, moreover, indubitable that iridectomy interposes a serious though not invincible obstacle in the way of a *prolapse of the iris*, and appreciably lessens the dangers of such a prolapse.

We have alluded in this connection only to the more common events that interfere with the healing process after extraction. There are others, more exceptional, however, in their character. To these belongs a disease which is clearly to be distinguished from genuine iritis—*iridophacitis* (observed six times in 1200 cases)—a state in which, while the inflammatory symptoms gradually increase, the whole capsular cavity becomes transformed—through hypertrophy of the intracapsular cells—into a bag of pus, while the iris at first plays but a subordinate part in the inflammation, and the ciliary region becomes only subsequently affected. In this category is, moreover, to be classed the affection to which I have given the name of *phagedenic wound-pustule*, and which may bring about a disastrous result as late as the third or even the fourth week; its occurrence, however, being fortunately even less frequent (four cases observed out of 1500). These processes are not, be it incidentally remarked, particularly affected by iridectomy. It accidentally happened that the most of our scanty observations were made on eyes on which a preparatory iridectomy had been done. Direct purulent infiltration of the vitreous, a thing that may develope itself after a loss of this humor or intra-ocular hæmorrhage, should also be reckoned in, were we making a careful statement of the things that might take place after an extraction, and not simply sketching their relation to iridectomy.

On a general review of the foregoing, it appears that iridectomy offers no protection against the occurrence of diffuse and partial suppuration, while on the other hand it does, in the case of the latter, go to ensure a more favorable course of things, and to a certain extent prevents the occurrence of iritis and prolapse of the iris. We may thence deduce the general principle that a portion of the dangers attendant on extraction are obviated by iridectomy, and that this operation is therefore to be employed where such dangers are imminent. As regards the particular indications, I stated several years ago (*Archive of Ophthalmology*, vol. ii., part 2, pp. 247–248, 1854) that iridectomy should always be done when the performance of the operation was attended with any difficulty; for instance, when, owing to the small size of the flap, to too small an opening in the capsule, or to too much adherence of the cortical substance, the lens did not slip easily out, but advanced with evident difficulty. Then I laid stress on iridectomy in cases where prolapse of the iris seemed probable, owing to unsatisfactory juxtaposition of the edges of the wound, to a tendency of the pupil to prolong itself in the direction of the wound. I do not, moreover, fail to perform iridectomy where entirely hard cataract exists in connection with a small pupil, difficult of dilatation; where cortical masses not yet completely ripe are with difficulty removed or have to be left behind, or where the former is true of a case of ripe cataract owing to the consistency and adherence of the cortical masses. Iridectomy, too, is to be advised in any case where the chance of a good union of the wound is small, because it may possibly contribute to a favorable result in the event of defined suppuration.

Is iridectomy always to be combined with extraction? Were the dangers of the operation actually diminished thereby, and were the procedure itself unaccompanied by any drawbacks, it would seem as though the question must be answered in the affirmative. Let candor, however, be scrupulously observed. The thing is by no means without its drawbacks. In the case of very restless patients its simultaneous performance meets with some obstacles and is attended with some anxiety; on neither of which points, however, am I inclined to lay as much stress as does my good friend Mooren, who has recently in a very meritorious manner drawn public attention to the combined procedure. An iridectomy done some time beforehand has the one decided disadvantage of subjecting the patients to a double operation, prolonging their stay in the hospital, and of sometimes, through delay, destroying their moral courage ; indeed, if the principle of a long interval be defended, it is often incompatible with external circumstances. More especially, however, is the fact undeniable that the vision of patients having a coloboma downwards and no accommodation, labors under some disadvantages— it not being necessary to regard cosmetic considerations with old people—when contrasted with cases where the pupil is central. These disadvantages apply less to the amount than to the distinctness of vision, and obtain particularly in instances where little irregularities of curvature or cloudy opacities of the cornea result in the vicinity of the wound after extraction, as not infrequently happens. These disadvantages are, to be sure, of minor consequence ; inasmuch, however, as iridectomy directly benefits only the minority of operative cases, and it is exceedingly probable that its performance under favorable circumstances is superfluous, I am unable, after weighing the pros and cons, to pronounce in favor of its general adoption in cases of extraction. It gives me pleasure, however, to see it done by others,* because an evident advantage accrues to science from its general application and a conscientious determination of results.

In the case of our patient, iridectomy is clearly indicated on the ground of probable suppuration of the flap.

It remains to decide whether iridectomy shall be done *at the same time* with the extraction, or shall *precede* it. The latter is, in my opinion, more judicious. Where the eye is unsteady its simultaneous performance is attended with certain practical difficulties or at least annoyances ; also, with the disadvantage of a subsequent small hæmorrhage into the anterior chamber, which although in itself insignificant, may embarrass the third and fourth steps of a modified linear

---

* The proposition to *always* combine iridectomy with extraction was made some time ago, as may be seen by referring to the following passage from my writings *(loc. cit.)* published nine years ago :—" Were any one to conceive the idea of making a pupil upwards a few weeks before extracting, as was in fact proposed to me by several, the only objection I should have to offer would be that in the infinitely greater number of cases the thing is unnecessary, and would be hardly compatible with the limited sojourn of the patients. On the other hand, such a procedure might be defensible on the ground of safety and as a prophylactic." With people who have only one eye, I have invariably practised this procedure for several years, for although, where other things are favorable, the danger of suppuration or iritis is not imminent, yet it must be taken into account in every extraction, and the lessened probability of it which is brought about by an iridectomy seems to me to outweigh the objections already referred to, especially when we consider the disastrous consequence of a want of success.

extraction, when iridectomy has been done as second act; moreover, while the recent formation of a coloboma certainly diminishes the tendency to a general iritis, the freshly cut edges of the coloboma do still incline to inflammation, resulting in union with the capsule, when irritated by the passage of the lens. These objections disappear when we find the coloboma ready made; the operation has then only its three usual steps, and is more quickly completed. I think that too short an *interval* should not be allowed to elapse between the iridectomy and the extraction. It is not advisable to do the operation within less than four, if possible six weeks of the other. Although the more evident signs of a tendency to irritability, consequent on iridectomy, may have disappeared, yet their traces may be observed in the redness of the parts met with after sleep, or in the minute infiltration in the neighboring portion of the cornea, revealed by oblique illumination; experience, moreover, shows that a too rapid succession of operations exerts a cumulative effect on circumstances disposing to inflammation, which, where the disposition to recovery is not a decided one, may give an impetus in the wrong direction. In short, if circumstances do not permit a longer interval than four weeks, I do the whole thing at one sitting, preferring this course, notwithstanding its disadvantages, to a double operation.

The poor patient whose case is in question is unable, owing to external circumstances, to lengthen her stay beyond what is absolutely necessary, still less to come back again. We have either, then, to do the iridectomy a week or ten days in advance, in contravention of the principle above established, or perform it simultaneously.

When, during the operation, unforeseen circumstances render necessary the performance of iridectomy, it forms of course the fourth step. Where the patient is restless, it is certainly attended with difficulty. Not infrequently too small a piece of iris is excised, the operator fearing to draw it too far forward, or dreading that an unexpected movement of the eye may give rise to dialysis or a loss of vitreous. Doing the operation, however, as the second step, immediately after the completion of the corneal section, involves much less difficulty, and is of course to be advised in all cases where, as in the present instance, the performance of iridectomy had been previously determined on.

January 5th.—Before proceeding to operate on the case of cataract recently referred to, I have the following remarks to make.

All so-called *preparatory treatment* is not only superfluous but mischievous, unless, indeed, there are special circumstances in the individual case requiring attention. The healing process, after extraction, goes on most favorably when the person operated on is in a high state of health, and is, so to speak, morally and physically well balanced. Even an active cathartic is to be shunned, the excitability of the patient being thereby increased. It is sufficient to induce a gentle evacuation by the use of castor oil, or some other mild laxative, the day before the operation. If we have a chance of previously watching the patient, it would be well to test the effect of a dose of morphine at least two days before the operation, in order to ascertain how the individual is affected by a drug we are so likely to subsequently employ, and which acts so differently in different cases.

A very important factor is the frame of mind of the patient, which may be such as to require encouragement. With many it is best to make no previous allusion to the operation. The course to be pursued must be left to the tact of the surgeon, who must lose no time in taking into account the imaginative tendencies of the patient. Experience shows that the patients who are free from care and full of hope do much better than those of excitable temperament and oppressed with anxious forebodings. During the operation, too, a word from the operator is, under many circumstances, of as much consequence as his manipulations. Avoid scolding or threatening, even when patients misbehave themselves, for such a course seldom fails to depress their spirits and paralyze their self-control. Jocular allusions, calculated to abstract their attention from the critical moment they are passing through, often have a good effect. It is a well-known rule not to let the patients wait long for the operation, but to perform it as soon as possible after it has once been decided on. Bad weather or indifferent light should not be allowed to weigh as much with us as depression of spirits on the part of the patient.

Is *atropine* to be instilled before the operation? Objections have been raised to this, particularly by English surgeons. It has been argued that the iris, being drawn back by the mydriatic, becomes thicker, and hence more liable to be injured by the cataract knife. It has been even argued that no real object is attained, inasmuch as the pupil contracts again after the evacuation of the aqueous humor;

and, finally, a fear has been expressed that the peripheric retraction of the iris may diminish the protection afforded the vitreous humor, and thus favor its escape. None of these objections appear to me well grounded. Even were the iris to increase in thickness as the pupil widened, it would be a matter of entire indifference; for if we take a case where extraction has been properly performed and examine the inner wound, it will be found that its arc nearly coincides with the edge of the enlarged pupil, or that its diameter barely exceeds that of the latter. It will therefore be found, in cases where mydriasis is complete, that the iris is either outside of the inner cut, or else that only a narrow border of pupil falls within it, and that this cannot possibly be so thick as to come in contact with the point of the knife, which is either close to the inner face of the cornea, or has attained to a corresponding point on the other side. It is true that the pupil contracts again after the escape of the aqueous. Leaving, however, out of sight the fact that the contraction itself is less than if no atropine has been employed, we should remember that our object is not so much to enlarge a firmly closed pupil—the influence of which in preventing a facile exit of the lens was formerly unduly feared—as to render more easy the completion of the section, during which, of course, the mydriasis persists. Finally, the objection that the use of atropine increases the chances of a loss of the vitreous is neither theoretically plausible nor confirmed by experience. I am accordingly disposed not only to dismiss the objections against, but to positively approve the use of atropine the evening before the operation. It is unquestionable that the completion of the section is thereby facilitated, especially when the anterior chamber is narrow, for the iris is thus entirely or nearly withdrawn from the region of the inner cut. It moreover strikes me as of importance that after the re-secretion of the aqueous humor the mydriasis in part returns, a fact that may be demonstrated by instilling atropine into the eye of an animal, then doing paracentesis and awaiting the refilling of the anterior chamber. This secondary mydriasis has some influence in averting a tendency to inflammation on the part of the iris. Finally, the paralysis of the ciliary muscle, persisting as it does after paracentesis, must be productive of good.

The operation itself has already been alluded to. We have yet to decide on the *direction of the section*. This paves the way for the general discussion as to whether extraction up or down offers the more advantages. Having once adopted the compressive bandage, the use of which brings with it the principal advantages of the upper section, I became more and more a partizan of extraction downwards, after having for six years practised exclusively the other method. My experience shows that an operation upwards and one downwards, each done according to rule, offer practically equal chances of success. Although, on the one hand, the average length of after-treatment of extraction upwards is three days less (twenty-one to twenty-four days' stay in the hospital), the percentage of prolapse of the iris, of iritis and of suppuration is the same; while, on the other hand, it must be conceded that partial suppuration does less damage in a case of lower than of upper section, the tendency to gravitate downwards being productive of evil results. It must further be alleged in support of the lower section that its employment renders it possible to do a greater number of operations strictly according to rule, the surgeon being less dependent on the docility of the patient, and even on that of the assistant. It is here in his power to prevent the flap from coming into rude contact with, or being everted by the lid, and to get rid of cortical substance without the introduction of instruments, &c., even when the patient has but little control over the movements of his eye. It is true that, with sufficient practice, the upper section gives very good results, even in cases where the eyes are deep seated, the patients unruly, &c.; still, the greater length of the operation and the danger of reversing the flap against the upper lid are disadvantages which, however skilfully they may be met, increase the amount of manipulation, interfere with the mechanical execution of the operation, and in a doubtful case may incline the scale in the wrong direction.

As is usually the case when the merits of different operations are under discussion, we have in the present case a variety of reasons for and against, the most of which experience has shown to be visionary. For instance, it has been alleged that vitreous was, by the law of gravity, more likely to be lost in extracting downwards. The reverse would be more correct. Vitreous escapes when the zonula or hyaloid are ruptured, a thing the less likely to happen in proportion as the exit of the lens and cortical substance is facilitated, which indubitably is true in the case of the lower section. It has been, moreover, asserted that corneal opacities, when they result, are productive of more disturbance in the case of the lower than of the upper section. Corneal opacities that amount to anything, however, only result from an exceptional condition of things, from prolapse of the iris or partial suppuration. The former would be accompanied by a dislocation of the pupil, which, in the general run of cases, had better take place above than below; while, in the latter, the pupil is either closed or contracted, and an iridectomy is indicated at any rate. It is also objected to the lower section that, if the edges of the wound are not in exact apposition, the border of the lower lid may get in the way; such a malposition, however, seldom occurs, and would not be allowed to remain. Whichever method were employed, a portion of the iris would have to be excised, and perhaps some vitreous evacuated, both of which things are more readily done in the case of the lower than of the upper section. Where the operation downwards has been done according to rule, the flap lies in such exact apposition that the lower lid causes no derangement of the parts, even when the gaze is suddenly directed downwards. Since commencing the systematic employment of the compressive bandage I do not, even where the eyes are unnaturally prominent, hesitate to extract downwards. I do not, however, give myself out as an opponent of extraction upwards. By so doing I should disparage the excellent results which many operators have thus obtained, and even the results of my own previous experience. I have already admitted that the average length of after-treatment is less in extraction upwards, owing to the fact that in a case of lower section, where no bandage has been used, the vicinity of the wound is more apt to be irritated by the action of the lids than when the cut is above and thus protected. I merely wished to draw attention to the fact that the much-vaunted superiority of the upper section is in many respects visionary, and that its real advantage, which consists in the uniform protection and pressure afforded by the upper lid to the wound during its first union, has been fully compensated by the introduction of the compressive bandage, and can no longer be used as an argument against the other method, the execution of which is more rapid, and in which we are less dependent on the patient. And if the results are, as I do not dispute, on the whole equally good, the preference must be given to extraction downwards on account of the greater facility with which it is performed. At all events, I advise the expert as well as the neophyte, in the case of a restless patient, to extract downwards.

In our own case we have, moreover, a special reason for choosing the lower section, viz., the iridectomy we are to interpolate as the second step. With the upper section the excision of a piece of iris, corresponding with the apex of the cut, is a very ticklish thing, unless we are aided by entire tranquillity on the part of the patient. Otherwise the excision must be lateral, and we gain only a partial advantage.

We shall, in accordance with our usual practice, operate on this patient *in bed*. Although this position is not entirely convenient for the operator, the patient enjoys the advantage of the greatest possible amount of muscular relaxation, and has afterwards to make no change. Surgeons who operate on their patients sitting are compelled, when anything out of the way occurs, to get them to bed during the operation, inasmuch as in certain contingencies—such as escape of the vitreous—the horizontal position is indispensable. As we are to deal with the right eye, the operator must sit behind the head of the bed, unless he happens by some freak of nature to be ambidexter, or else during his medical education was influenced by a now exploded ophthalmic hobby and worked up a partial ambidexterity, for worked up it must be. The choice is free to all; still, being sworn foes to all surgical coquetry, we would remind those who plume themselves on ambidexterity that the creation of practical difficulties, when such might with entire propriety be avoided, is less a merit than an indiscretion. It is not likely that any surgeon would acquire a title to dexterity because he amputated one leg with the right hand and the other with the left, instead of so varying his position as to do both with the right hand. The same holds with regard to ophthalmic operations, provided the principle of operating on the patient in a recumbent position be once adopted.

January 5th, P.M.—There is but little to say with regard to the operation which has just been performed. The measurement of the cornea showing its diameter to be but small, the cataract knife had to be introduced close to the scleral edge. Under the circumstances it was only thus possible to form a cut of sufficient size; in cases, however, where the cornea is larger it is practicable to make both puncture and counter-puncture at the distance of a millimetre from the scleral edge. When the points of entrance and emergence are taken close to the scleral edge and the cut rounded off to correspond, we are apt, as in the present case, to get a small flap of conjunctiva, to the existence of which we are neither disposed to attribute unpleasant consequences nor attach a particularly favorable significance. Both views have, however, been held, the last, as is well known, having of late been strongly insisted on by an eminent practitioner, who regards the conjunctival flap as a strong safeguard against suppuration of the wound, and insists on its regular formation. When the conjunctiva is cut we do not regard it advisable to form a long and narrow flap, because that is apt to bleed and may prevent apposition of parts. It is better to turn the knife at a right angle to its previous position and cut directly out.

Just before completing the flap, the fixation forceps, the use of which in cases of flap-extraction may be conscientiously recommended, were laid aside, and the upper lid suffered to drop completely down. Manifest advantages attend the latter manœuvre, which is, of course, only practicable with the lower section. If the cut has approached within about 1''' of its completion, its form is already a settled thing and no particular results would accrue from turning the knife more or less forwards. No further reason, therefore, exists for requiring as much of a palpebral aperture as is usually sought. The assistant should let go of the lid. The operator, who has laid down the fixation forceps and therefore has his other hand at his disposal, should use it for the purpose of gentle traction downwards on the cheek, in order to complete the cut while withdrawing the knife, without wounding the edge of the lid, and yet under the same circumstances as if both were closed. The palpebral pressure is thus, at this critical moment, reduced to a minimum, and protrusion of the iris, sudden loss of the aqueous humor, or even escape of the vitreous, are less liable to occur with unruly patients.

Collapse of the cornea, which the deep-set position of the eyes and the diameter of the cornea had led us to predict, took place after the first step (of the operation). It became strongly marked after the completion of the other three steps, concerning the execution of which I have nothing to add. The cornea finally nestled down in plaits, the edges of the wound, however, being in entirely good apposition, as was tested by restoring with the finger some degree of tension to the ball.

Immediately after the operation the *compressive bandage*, in common use at my infirmary, was applied. The orbital hollow is first evenly packed with charpie, which has been picked over and put together in the form of small tufts, the whole being secured by a single turn of a snug-fitting flannel bandage passing over one eye. This is held in place by another single turn around the forehead, the first half of which comes before, the other half after the turn passing over the eye. The middle portion of the bandage passing over the eye is knit of cotton and not of flannel. Special stress is to be laid on the proper management of this bandage* in cases of collapse of the cornea. Tolerably firm pressure must be made during the first

---

* I am preparing for the Archive an article on the subject of this bandage. [This has already appeared. See Archiv für Ophthalmologie, Bd. ix. Abth. 2, S. 111.—*Translator.*] I would only here observe that to Sichel particularly, of modern ophthalmologists, is due the credit of the introduction into practice of the compressive bandage after extraction (since 1842, see Gaz. des Hopit., 1853, No. 54),'and that I was incited by the personal solicitation of Sichel to devote time to the study of this important subject. A letter of mine on the subject of the compressive and constrictive bandages may be found in the "Manual of General Ophthalmology," by Seitz and Zehender, p. 425 et seq., Erlangen, 1861.

few hours, and then gradually relaxed in order not to hinder the escape of liquid secretions.

January 6th, A.M.—Fifteen hours have passed since the operation without any complaint. Even the pain about the cut, which succeeds the operation, was totally absent. Is this to be construed as influencing for the better our unfavorable prognosis ? Not in the least. In marasmic eyes we not infrequently meet with a similar amount of sluggishness, lasting twelve, sixteen or eighteen hours, and then yielding to symptoms of diffuse or circumscribed suppuration. In looking through my records for the after-history of extraction done on marasmic eyes, I find that where there is absolute insensibility the prognosis is less satisfactory than in cases where a few hours after the operation a certain amount of pain developes itself in the wound, accompanied by slight swelling of the lid, reaching its height in from four to six hours and then disappearing.*

6th, P.M.—The apprehensions expressed this morning have since unfortunately become realized. Shortly after the morning visit, about sixteen hours after the operation, the patient experienced a sensation in the eye, at first uncomfortable and then painful, and noticed an increased flow of tears. The bandage was removed eighteen hours after the operation. The nethermost layers of charpie were well soaked with a clear fluid. Since then both the swelling and the secretion have steadily increased. We now find, on opening the bandage, the lid already considerably tumefied, the folds broader, the deep hollow between the edge of the orbit and the bulb sensibly filled out in comparison with the other side, the whole superficies of the lid increased in volume. The furrow just above the internal palpebral ligament shows the characteristic swelling, œdematous and with a dash of redness. The flow of tears not only did not diminish during the morning, but has now become more and more mixed with a muco-purulent secretion. Although the bandage has been on only four hours, the whole lower layer of charpie is now soaked with pus. On removing the whole of the bandage several drops of a yellow secretion were observed to adhere to the edges of the lids. There

* This applies, of course, to marasmic eyes alone. In other cases, where the operation has gone off in the usual manner, an entire absence of subsequent pain from the very first is highly favorable; although it is an acknowledged fact that pain occurring within eight hours of the operation need cause much less anxiety and yields much more readily to treatment than when it comes on between twelve and thirty hours afterwards.

is, to be sure, somewhat less pain, a circumstance which in no wise diminishes our apprehensions, inasmuch as an abatement of the pain is often noticed as the suppuration becomes developed.

To what shall we *attribute the symptoms?* Without doubt to a suppurative process of the cornea. Whether diffuse or circumscribed cannot be decided till the lids are opened. A simple prolapse of the iris is out of the question. It might, to be sure, cause a moderate swelling of the lid and an increased flow of tears; the proof of its existence would be, however, the fact that the tears remained clear, and, even when persistent, were very seldom mixed with masses of mucus; the occurrence, too, of a prolapse of the iris after an iridectomy and a properly practiced incision, is very rare. It consists, too, in the outset, of a slight involvement of the iris, which, in the course of time, gradually becomes more prominent and gives rise to symptoms. Still less can the present case be one of genuine iritis. Leaving out of sight the fact that this need hardly be apprehended in a case where the cataract was easily and entirely removed and iridectomy had been done, its existence would be indicated by neither the period, the symptoms, or especially the abundant secretion.

Can there, under existing circumstances, be any harm in *opening the lids?* Although strongly disposed to postpone opening the lids till the fourth or fifth day in a case where all has gone along quietly, because a different course would be certainly unnecessary and possibly prejudicial, I see no objection, where things are going wrong, to gain a knowledge of the facts by a careful examination of the eye. Granting that we have here a case of general suppuration, we should then relinquish all hope and adapt our treatment to the general condition of the patient, studying her comfort the while. If, on the contrary, the case be one of circumscribed suppuration of the wound, preventing its closing, some hope would still remain, and we should have to be guided by a careful study of the healing tendencies.

On proceeding, therefore, to open the eyes, several drops of a fluid composed of tears mixed with pus escaped from the conjunctival sac. The conjunctiva bulbi is tolerably reddened and pretty strongly chemotic. The entire corneal wound has become infiltrated with opaque, yellow matter to the extent of nearly 1''' and evidently throughout its entire thickness. The whole corneal flap moreover has a yellowish, sodden appearance. Through its upper third alone is the iris visible and the reëstablishment of the anterior chamber evident. The corneal wound does not actually gape; a study, however, of this stage in similar cases leads us to infer that the cut edges are not in contact throughout their entire thickness. The patient has satisfactory quantitative perception of light.

How are we to regard this process? From which *tissue* does the suppuration proceed? There may those be still found who regard the iris as the source of the affection. But my anatomical and clinical studies have brought me to an opposite conclusion. When more extensive and spontaneous purulent infiltrations of the cornea show a tendency to extend to the epithelial layer of the membrane of Descemet and thence to the iris, much more would this tendency to diffusion be present in the case of a penetrating wound and violent suppuration. There is nothing surprising, then, in the suppuration of the wound being closely followed by turbidity of the aqueous and secondary iritis, the latter of which more readily attracts attention than the original affection, where cicatrization has already commenced. Nor has the supposition that the visible symptoms proceed from a suppurative cyclitis or choroiditis, any clinical foundation. It is true that such results may follow extraction in rare and exceptional cases. Loss of vitreous, for example, may be immediately succeeded by a purulent infiltration of the deeper structures; in an ordinary case of suppuration, however, like the present, the deeper membranes become involved either by an extension of the suppurative affection of the cornea in the shape of a diffused purulent choroiditis or panophthalmitis, or else the results of the secondary iritis develope themselves insidiously, the products of inflammation on the posterior face of the iris attaining a complete union with the capsular cavity, and the whole resulting in a chronic cyclitis and a consequent atrophy of the bulb. A trial of the amount of perception of light enables us in each case to form an opinion. Even in cases of diffuse suppuration of the cornea we find, during the stage of the so-called peripheric abscess, both perception of light and field of vision normal, so long as no signs of deep-seated suppuration, such as rigidity and protrusion of the bulb, are present. This could not possibly be so, were the purulent affection of the cornea but an evidence of suppurative choroiditis. An anatomical investigation proves decisive, and this we once had an opportunity of making in an entirely typical case of suppuration of the cornea (peripheric abscess, coming on thirty-six hours after extraction, and described at length by Dr. Schweigger). The deeper structures were here found entirely normal.

Is the corneal suppuration to be regarded in the case of our patient as already *diffuse* (i. e., past aid) or *circumscribed.* Although the transparency of the greater part of the cornea has become impaired, yet it is only within the narrow streak following the edge of the wound that the infiltration has taken on the character of destructive suppuration. The remaining semi-transparent infiltration is at present to be considered as a swelling of the parenchyma, capable of retrogression. Moreover, a careful examination of the upper half of the cornea reveals the entire absence of that ominous circular infiltration of pus, which, when it entirely surrounds the cornea, is a certain presage of necrosis and slough, and the proper pathognomonic mark of diffuse corneal suppuration. If, however, the case at the present time be one of circumscribed suppuration, or suppuration of the wound, it is highly probable that it will take on the diffuse form. Hardly twenty-eight hours have elapsed since the operation, the symptoms have been continually on the increase, the morbid process has its seat in a part which was known in advance as offering a poor field for action and inclined to suppurative processes; every circumstance is therefore unfavorable. Should diffuse suppuration not occur, hopes may yet be entertained; the processes thus set up, however, go to endanger the ultimate result, although here the existence of the coloboma is much in our favor.

What can we do that will tend to limit the suppuration? Shall ice-cold applications be made? While they are of doubtful utility in a prophylactic point of view after an operation, they are to be decidedly condemned when suppuration has once commenced. This is eminently true of marasmic eyes. The application of cold does but accelerate the advance of the process most to be feared.

Nor can I, under existing circumstances, recommend the application of leeches in the vicinity of the eye. Just in proportion as their use is beneficial in cases where the wound has healed, but where symptoms of irritation appear along its course or iritis comes on, so is their employment attended by danger during the early part of the healing process. Antiphlogistic treatment under such circumstances owes its continued employment to the erroneous idea that whatever happens after an operation depends necessarily on an undue amount of inflammation. When a marasmic patient has a wound that threatens to suppurate, no surgeon would think of applying leeches in its vicinity. And does not the same principle hold good of a part the already feeble nutrition of which has been impaired by an extensive incision? The afflux of blood, induced by the suction of the leeches, naturally causes a more abundant inundation of the infiltrated parts, and the affair terminates unfavorably before the occurrence of the period of remission, from which good might otherwise be derived. It was sad experience, however, and not the theoretical considerations which more tardily matured, that has led me to abstain from the application of leeches in the vicinity of the eye the first three days after an extraction.

Shall venesection be performed? I admit that this method is free from the disadvantages pertaining to leeches, that time is thus often gained when the symptoms are threatening, and the affection thus carried forward into a more propitious phase. But with marasmic individuals it might have a deleterious effect upon the general health, while the chances of its furthering the healing process are more than problematical. Finally, when it is employed, it should be at the inception of the symptoms and not after suppuration has declared itself.

What remains to be done if thus much be rejected? For a considerable time I have been in the habit of employing the so-called *constrictive bandage* in cases of circumscribed suppuration, and alternating it with warm aromatic applications when treating marasmic individuals. The former of these applications I cannot praise too highly. I have, within the last four months, seen its use followed by remarkable results in three cases which were running an anomalous course, and where some of the symptoms were more developed than in the present case.* In applying the constrictive bandage the orbit is packed in almost the same manner, only somewhat more plentifully, as when the compressive is used; the whole being firmly secured by a flannel roller passed three times over the eye in question.† One shrinks at first from placing such a bandage on an eye

---

* In one of these cases the commencement of a peripheric abscess existed, two thirds of the corneal edge remaining, it is true, free; in spite of which the eye was saved, and hopes are entertained from a future iridectomy.

† I shall take advantage of some proper opportunity to advert to some points connected with this bandage; suffice it here to say that it has an entirely different effect from the pressure bandage, inasmuch as it not only actually supports the eye, but firmly presses the flap.

where the lid is swollen and an active suppuration going on; experience shows, however, that when the proper indications are met, the patients find the application comfortable, and that swelling and suppuration diminish to an extent that, under such desperate circumstances, could have been expected with no other method. Such a bandage is ordered for this patient, with directions to have it changed every three hours, subsequently at longer intervals, in case the suppuration shows the desired diminution. Between the applications of the bandage, camomile fomentations at a temperature of 95° are to be used on the lids, it being expected that they will contribute to the termination of the suppurative process. The diet is not to be lowered; the patient, on the contrary, is to be plied with bouillon and to drink milk.

January 7th.—During the last eighteen hours the bandage has been applied three times. The female attendant states that the first time it was removed there was a decided, and the second a notable diminution in both the swelling and discharge. On removing it now only the under surface of the lowest layer of charpie is found permeated by a purulent discharge, which is certainly infinitely less abundant than yesterday. The secretion begins to dry along the palpebral aperture, a fact which also denotes a diminution in quantity. The lid is still swollen, though unmistakably less, its folds are becoming less broad, more prominent, while here and there is seen a tendency to wrinkle. On opening the eye, as much improvement is seen as could have been expected in so short a time. The purulent infiltration at the lower part of the wound is beginning in some places to come away in the shape of a muco-purulent coating. The remaining corneal opacity extends as high up as yesterday; has changed its color, however, from yellowish to gray. Incisions are made in what chemosis exists; the constrictive bandage is directed to be left on from six to eight hours, and camomile fomentations to be used but half an hour at a time in the interval.

January 9th.—During the last two days improvement has been more and more decided. The bandage was finally left on twelve hours together, and on its removal the lower layer of charpie was found dry, and the bit of linen covering the lids was all that the secretion caused to adhere; this was, moreover, dry. The swelling of the lids had ceased, and the chemosis disappeared. By the aid of the forceps a species of muco-purulent slough could be picked from

12

the wound, and its removal brought into view the purulent infiltration of the edge of the wound, still in existence, but tending to consolidation. The infiltration of the flap had hardly a tinge of yellow in its appearance, but looked a good, healthy gray. Through its upper and already transparent edge could be seen the border of the discolored pupil (iritis propagata). The constrictive bandage was ordered to be continued and atropine to be instilled; the warm fomentations were omitted.

January 20th.—The progress of the case has been as favorable as could possibly have been anticipated. The purulent infiltration in the vicinity of the wound has become more and more consolidated, and will leave a cicatrix about $\frac{3}{4}'''$ in breadth. A tissue in process of organization may be seen to extend from the wound to the pupil, indicating the course of the previous suppuration. The effect of the atropine may be seen in an enlargement of the pupil upwards, which is but very lightly filled with recently developed material, and allows the patient to count fingers held close to her. Although the artificial pupil is entirely filled and contracted, it must be allowed that its existence has materially contributed to the relatively favorable course of the transplanted iritis. Without it we should probably have had a suppurative and total fusion between the edge of the pupil and the capsular cavity, an accumulation of exudation on the posterior wall and cyclitis at a time when any fresh operative interference was not to be lightly undertaken. In contrast with this we have the iris in an entirely normal plane, nowhere bellying forward, but slightly disorganized, free from large vessels—in short, the ciliary region evidently uninvolved.

That the eye did not fall a prey to diffuse suppuration, nay more, that a favorable *terrain* has been gained for a subsequent iridectomy, we are indubitably to ascribe to the use of the *constrictive bandage*, kept up for five days after the date of the last record. In this therapeutical conclusion I shall be confirmed by every colleague whose own experience has enabled him to gain an insight into such incidents as have been described, occurring so soon after the operation of extraction. Had it been possible for me to have seen the patient on the 6th, about eight hours before the evening visit, and thus previous to the more abundant suppuration, and at that time to have ordered the constrictive bandage, it is probable that the use of this, under more favorable circumstances, would have produced more brilliant results.

ERRATA.

P. 17, 19th line from top, *for* intercranial *read* intracranial.

P. 21, 17th line from bottom, *for* " that an affection of the fasciculus lateralis " and remainder of sentence, *read* " that an affection on the one side of the fasciculus lateralis, on the other of the fasciculus cruciatus alone, and not of the contiguous fibres, should be in question."

P. 24, 9th line from top, *for* altered demeanor *read* a sense of confusion.

P. 28, 1st line, *for* employée *read* employé.

www.ingramcontent.com/pod-product-compliance
Lightning Source LLC
Chambersburg PA
CBHW021954190326
41519CB00009B/1254

9 783337 840327